GREEN HAWAI'I

A Guide to a Sustainable and Energy Efficient Home

KEVIN J. WHITTON

*James,
Keep it green!*

Mutual Publishing

Copyright © 2008 by Mutual Publishing, LLC

All rights reserved. No part of this book may be reproduced in any form or by any electronic or mechanical means, including information storage and retrieval devices or systems without prior written permission from the publisher, except that brief passages may be quoted for reviews.

Library of Congress Cataloging-in-Publication Data

Whitton, Kevin J.
  Green Hawai'i : a guide to a sustainable and energy efficient home / Kevin J. Whitton.
     p. cm.
  Includes bibliographical references.
  ISBN 1-56647-861-8 (softcover)
  1. Sustainable buildings--Design and construction. 2. Buildings--Energy conservation. 3. Sustainable architecture. I. Title.
  TH880.W45 2008
  644.09969--dc22
                              2008007888

ISBN-10: 1-56647-861-8
ISBN-13: 978-1-56647-861-8

Images on pp 2, 3, and 13 courtesy of HECO
Images on pp 14, 18, 29, 30, 32, 33, 107, and 108 courtesy of Dean Masai with the DBEDT Energy, Resources and Technology Division
Images on pp i, 6, 7, 8, 44, 77, 79, 80, 82, and 83 courtesy of John Harrison
All other images by Kevin and Michelle Whitton

Design by Emily R. Lee

First Printing, September 2008

Mutual Publishing, LLC
1215 Center Street, Suite 210
Honolulu, Hawai'i 96816
Ph: 808-732-1709 / Fax: 808-734-4094
E-mail: info@mutualpublishing.com
www.mutualpublishing.com

Printed in China

contents

v	Acknowledgments
vi	Foreword
xii	Introduction

In the Home
3	one: energy
27	two: design
39	three: indoor water conservation

In the Yard
47	four: landscaping
77	five: remodeling

In the Community
87	six: consuming
93	seven: recycling
103	eight: volunteering
107	nine: comfortable living

112	References and Further Reading
114	About the Author

Acknowledgments

The enjoyment and preservation of nature have always been of paramount importance to me. I would like to thank Jane Gillespie at Mutual Publishing for presenting me with the opportunity to write this book, which has not only expanded my knowledge of and interest in this subject, but also introduced me to a community of conscientious and passionate people who have dedicated their lives to the betterment of Hawai'i. The courtesy, willingness to help, and aloha shown to me by the all the people I spoke with in conjunction with this book cannot go unmentioned.

I relied on several professionals and authorities on the different facets of energy efficiency and sustainability in my research for this book. Dr. John Harrison, PV guru and environmental coordinator at the University of Hawai'i at Mānoa's Environmental Center; Ron Richmond, solar hot water systems expert with the Hawai'i Solar Energy Association; Steven Meder, professor at UH Mānoa's School of Architecture and director of the Center for Smart Building and Community Design; and Ray Heitzman, owner of Ray's Solar Fans and independent solar contractor, were instrumental in the energy-related sections of the book. Jill Laughlin, education and volunteer programs coordinator at Lyon Arboretum; Kevin Eckert, ISA-certified arborist and consulting arborist for Arbor Global; and Leland Miyano, naturalist and landscape architect, donated time, energy, and knowledge covering the outdoor topics. Dean Masai with the Department of Business, Economic Development and Tourism (DBEDT), Energy, Resources and Technology Division, rounded up some drawings and images for the book; and a special thanks goes to Janeen Loose, corporate communications for Hawaiian Electric Company (HECO), who provided me with information, contacts, images, and the true spirit of aloha, despite all my emails and calls. Without these individuals and the kind and generous people who opened their homes to me as examples of sustainability, this book would not be possible.

Foreword

Climate change is upon us. The Intergovernmental Panel on Climate Change (IPCC), comprised of more than 2,500 scientists and other experts from countries around the world was co-recipient of the 2007 Nobel Peace prize, along with former Vice President Al Gore. In awarding this year's prize, the Nobel committee recognized the valuable contributions of a growing community to draw the world's attention to the pressing challenge of a changing climate. In a series of four publications this year, the IPCC noted that climate change is "unequivocal" and that human-induced warming has consequences, including species migration, glacial retreat, thawing of arctic permafrost, and increasing rates of sea level rise. The panel's concluding Synthesis Report enumerates the multiple observations of climate change and notes that

> warming could lead to some impacts that are abrupt or irreversible, depending on the rate and magnitude of the climate change.

What are we going to do about climate change? The challenges we face certainly are daunting; the prospect of dire consequences emphasizes the need for broad and enthusiastic adoption of sustainable living practices. The word "sustain" is primarily defined as "to keep in existence." Thus, the concept of sustainable development, as proposed in 1987 by the Brundtland Report, is

Wai'anae mountains.

> development that meets the needs of the present without compromising the ability of future generations to meet their own needs.

Historical examples of human impacts on the natural world and their repercussions are abundant. Indeed, the oldest recorded human story, the Epic of Gilgamesh, carved in Sumerian tablets dating from 4,000 years before the present (YBP) and unearthed in the cradle of human civilization, is ultimately a tale of struggle for dominion of Man over

Nature. King Gilgamesh seeks to tame wild Enkidu, lord of the beasts, and after a struggle, they make peace, and Enkidu is brought into the realm of humanity. The two protagonists then proceed to again confront and subdue raw Nature in the form of the demon Humbaba who guards the sacred cedar forest. At the end of the conflicts, Enkidu and Humbaba are dead, the sacred cedar forest is destroyed, and Gilgamesh is left contemplating the prophetic apportionment of death to mankind and eternal life to the gods.

Early city-states in the Indus Valley, including the great civilizations of Harrapa and Mohenjo-daro, flourished at the time the Gilgamesh epic, which had been passed orally from generation to generation, was first recorded. Ironically, the fate of these cities and their civilizations paralleled that of the humans symbolized in the epic. Extensive population growth and unsustainable irrigation practices may ultimately have led to an inability of the communities to feed their people because of changing water courses, increasing salinity of irrigation water, and eventual aridity of the land.

Interest in the relationship between the natural world and human well-being evolved during the pre-Socratic Greek era among the Sophists, who questioned the prevailing belief system that attributed all human fortunes and misfortunes to the Greek pantheon of deities. During the period between 2.6 and 2.3 thousand YBP, recognition of observation-based natural phenomena became more widespread among philosophers such as Hippocrates. One of his most famous treatises, *On Airs, Waters and Places*, attributes disease to an "unhealthy site" rather than punishment meted out by spiteful gods. More generally, Hippocrates linked air quality and climate to health and national character. Theophrastus, a student of Aristotle often referred to as the father of botany, noted how the clearing of forests for agriculture led to changes in the local climate.

In the New World, the pre-Columbian Mayan culture (2.0 to 1.0 thousand YBP) achieved impressive civil engineering feats. In addition to architectural accomplishments throughout their realm, the Mayans

Pipeline.

Lili'uokalani Botanical Gardens.

drained marshes, cleared and terraced hillsides, and undertook measures to prevent widespread erosion and expand lands for agriculture. However, the decline of the Mayan civilization was underway by 1100 YBP, most likely due to excessive stress on land used for food production to meet the needs of the expanded population.

Sensitivity of human communities to relatively small (<1-2°C) temperature changes from global interglacial conditions is apparent in the consequences of the period of cooling that occurred between 800 and 200 YBP, often referred to as the Little Ice Age. Severe winters during this era led to shorter growing seasons and famines, in turn resulting in the retreat of vineyards and grain production from northern Europe. Disease, famine, and abandonment of settlements induced widespread migration of populations from Russia and Scandinavia to more western and southern regions. In turn, these migrations produced a permanent change in the cultural, demographic, and political distribution of European civilization. Among other effects, the combination of growing population, increased deforestation, and more severe winters led to a widespread shift in heating fuel in Britain from wood to coal. By the mid thirteenth century, air pollution in London was so severe that commissions were established by Henry III to consider the problem of air quality, and in 1273 an ordinance was enacted prohibiting the use of coal in London because it was prejudicial to health. Henry's successor, Edward I, carried sanctions a step further, imposing a penalty of death on anyone burning coal. Necessity, as well as better fireplace designs, eventually led to renewed use of coal for heating. As a result, the problem of air pollution due to fossil fuel combustion not only predated the Industrial Revolution by centuries but also persists to this day.

In the nineteenth century, an increasingly detailed understanding of dynamic interactions between the atmosphere and surface temperature began to emerge as a result of studies by Jean Baptiste Joseph Fourier, best known for the mathematical formulations that bear his name. In 1824, Fourier discovered that atmospheric gases may increase surface temperatures, comparing atmospheric heating and the action of glass in a greenhouse. In 1861, the Irish physicist John Tyndall greatly expanded on Fourier's observations with direct measurements of the heat absorptive powers of various atmospheric gases. In particular, Tyndall's work laid the foundation for the later work of chemist Svante Arrhenius, which led to his postulation of the greenhouse effect. However, residing as he did in Sweden, Arrhenius felt that global warming might not be such a bad idea:

Looking north from Makapu'u.

> By the influence of the increasing percentage of carbonic acid [carbon dioxide] in the atmosphere, we may hope to enjoy ages with more equitable and better climates, especially as regards to the colder regions of the Earth.

With the benefit of better knowledge and more sensitive technologies with which to monitor the extraordinarily complex planetary systems whose balance our activities have upset, we are beginning to appreciate the perilous transformations we have engendered. Along with this understanding has come an appreciation for the urgency with which we must effect changes in the direction of sustainability, as well as the magnitude of the task before us. In an essay written in 1989, William Ruckleshaus, the first administrator of the Environmental Protection Agency asked,

> Can we move nations and people in the direction of sustainability? Such a move would be a modification of society comparable in scale to only two other changes: the agricultural revolution of the late Neolithic and the Industrial Revolution of the past two centuries.

Ruckleshaus's allusion to the agricultural and Industrial revolutions helps to underscore not only the immensity of the task facing our modern world, but also the time scale over which it is likely any effective remedial action will occur. Most planning is undertaken within a limited time horizon, often defined politically in terms of four years or less. In retrospect, the great historical transformations of the world's economic and social institutions have occurred over intervals of centuries. The challenge of transforming our energy economy and lifestyle from traditional yet unsustainable conventions to renewable self-reliance is daunting, but the alternative—complacence—does a great disservice to future generations.

Some have drawn analogies between the obstacles we confront in achieving this new paradigm and those faced by proponents of the Apollo program, which sought to expand the tangible reach of humanity to our moon's surface. Certainly, the momentum of national commitment must be at least equivalent, but two substantive factors bring this leap of achievement more readily within our reach.

First is the expansive, cooperative base upon which the endeavor rests. Unlike the elite community of technical and scientific innovators upon which the Apollo program relied, the sustainability transformation will arise from the collective efforts of all members of our community. It is gratifying to observe simple choices of efficiency over waste that are now evident throughout our world, and gaining popularity every day.

Sunset.

Second, and more fundamental to our success, the challenges of space travel and survival on the surface of the moon required invention and innovation, leading to the new technologies and the new science necessary to accomplish the Apollo triumphs. In contrast, we already possess virtually all of the technological and scientific know-how to achieve sustainability. No daunting knowledge barriers bar entrance to the sustainable future we envision.

Certainly, the transformation of our society to an ethic of sustainability will require our collective will and efforts. It also will require inspired leadership from informed citizens who are determined to make sustainable choices. Kevin Whitton has provided valuable guidance in this book, collecting principles and examples of sustainability that are laid out logically and in an easily accessible format. Part 1 offers comprehensive treatments of home energy and resource management, including both technical and practical design alternatives, which, carried to full implementation, offer a pathway to self-sufficiency and the potential to return excess energy to the community. In Part 2, Whitton addresses the crucial consideration of the ecosystem immediately surrounding the home, pointing out the key interactions between vegetation and microclimate, as well as principles of structure and materials that help to maximize energy and resource efficiency. The concluding chapters expand the horizons of personal contributions to sustainability into the immediate community, with valuable guides to the choices and actions that individuals can embrace in the pursuit of a more benign planetary footprint.

Hibiscus.

As one who has professionally and personally embraced sustainability as a survival necessity, rather than merely the politically correct thing to do, I have felt privileged to contribute to this valuable book. To those readers looking for encouragement along the path to sustainability, this is a very good place from which to embark. Although we may look to our government leaders for initiatives to reverse destructive patterns of behavior, it is ultimately the sum of individual choices, commitments, and actions that will light the way to a sustainable future. And while it truly is cool to be the only house in the neighborhood that has lights, music, and cold beer when the grid power goes down, it will be much cooler to look out at an entire neighborhood of distributed generation and contemplate the tons of $CO2$ that each sustainable house in the community displaces from our skies.

John T. Harrison, PhD
Environmental Coordinator (retired)
University of Hawai'i at Mānoa
Honolulu, Hawai'i
January 1, 2008

Introduction

Lake Loko Waimaluhia, Hoʻomaluhia Botanical Garden.

When my wife and I moved to Hawaiʻi, we bought a small home on the windward side of Oʻahu, close enough to town for work, but far enough away to capture the natural tropical beauty of the island. We are close to the ocean, where the trade winds cool our simple, 1,000-square-foot rectangular house and keep me busy in the garden picking up the potted plants that every so often lose their footing in a big gust and topple over.

Our house was built in 1971 with single-wall construction, jalousie windows, and no insulation in the attic. We don't have air conditioning, but most of our neighbors do—the boxy single-room air conditioning units that hang out of the window frame like a noisy white tumor, rusting on the side of the house. There are no big trees near our house, only two neighboring clumps of areca palm. The people who lived in the house before us installed metal awnings over the two west-facing windows in the front of the house to block the hot afternoon sun. The exterior walls are painted a light tan color and we have ceiling fans in every room. We have great natural ventilation because there are windows in every room and our home stays relatively cool with the passing breeze. However, when light and variable winds are predominant—or even worse, when there's no wind at all—our house transforms into a sweltering oven. We have a small, 30-gallon electric hot water heater that serves both of our needs for hot showers and dishwater. We are active, outdoors people and do not have a TV, a microwave, or bad habits like leaving the lights on when we are not home.

That's why it came as such a shock when I opened our first electricity bill and stared helplessly at the $95.54 that we owed for our monthly energy consumption. That total wasn't elevated due to activation fees or for an additional pro-rated timeframe; $95 continues to be our average month-to-month bill.

I pored over the statement for some type of clerical mistake, vaguely understanding the breakdown of taxes and kilowatt-hours, and wondered how one month in our lives could consume $95 worth of electricity. I called HECO, asking if there was some mistake. The customer service representative politely told me that $95 was pretty reasonable for the area and that electric bills don't get much lower.

I stopped turning on the porch light at night, I turned off the computer every time I was finished using it. I made sure the stereo and kitchen appliances like the blender, coffee maker, and toaster oven were unplugged when not in use. This hardly affected our monthly bill. I threw up my hands in defeat, assuming that if this was life in paradise and $95 dollars a month was what it would cost for light, so be it.

I was just your average energy consumer, inextricably tied to the grid for power, for food, for what seemed like life. I concentrated on paying the mortgage, doing good things, and enjoying life. I thought I was doing my part by turning the lights off when I wasn't in the room. But then I started doing research for this book, *Green Hawai'i*, and I realized how easy it was to make a change for the better and how that change could have real, tangible consequences that would reverberate into the community and into the world.

I don't have a lot of money; I can't afford a photovoltaic system for my home at this point in my life, but I have switched all my incandescent light bulbs to energy-efficient CFL bulbs. I've insulated my hot water tank until I can afford to write the check for a new solar hot-water system, and I've planted shrubs and small trees around the house to help shade the thin walls from the sun. My wife and I practice water conservation, indoors and outside, and donate our recyclables to the local school up the street.

Right now, that is our contribution, and it matters. Every contribution matters. No effort should be taken for granted, no matter how seemingly small or inconsequential. As I educate my family and yours as well, we can add to the measures that we have already taken to make our homes "green" homes. And as people get on board and make environmentally friendly and sustainable choices, one at a time, together we can make our homes and Hawai'i a greener place to live.

The Islands provide us with food, water, and a quality of living that is nonpareil in this world. Let us look to our constructed homes as well as our Island home with respect and understanding, and take the necessary actions so that future generations can enjoy Hawai'i's natural beauty and splendor.

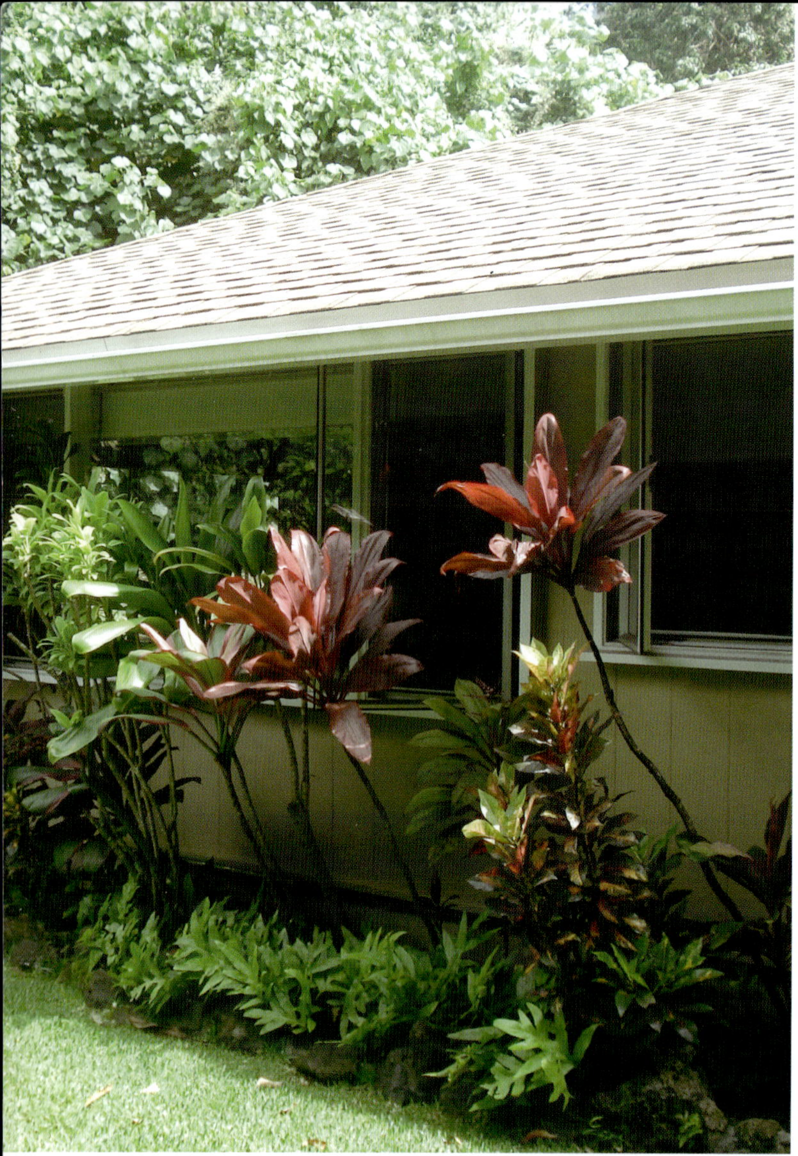
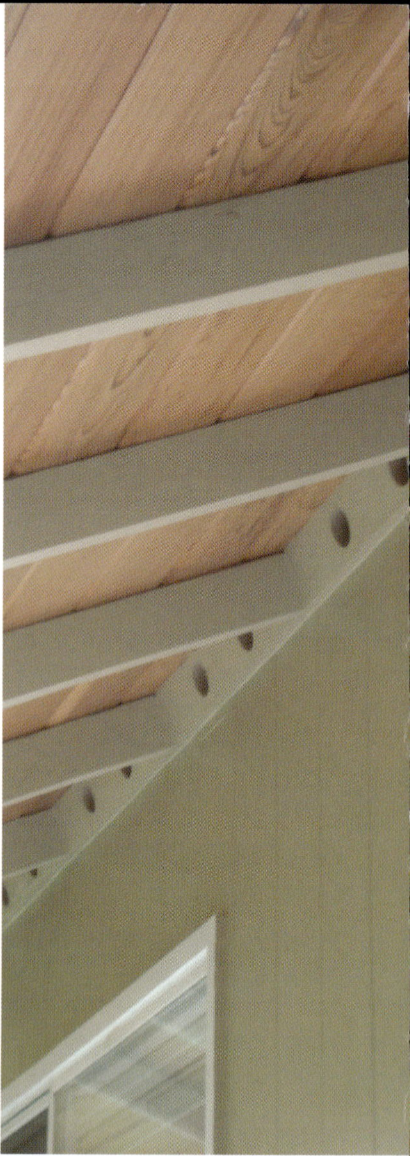
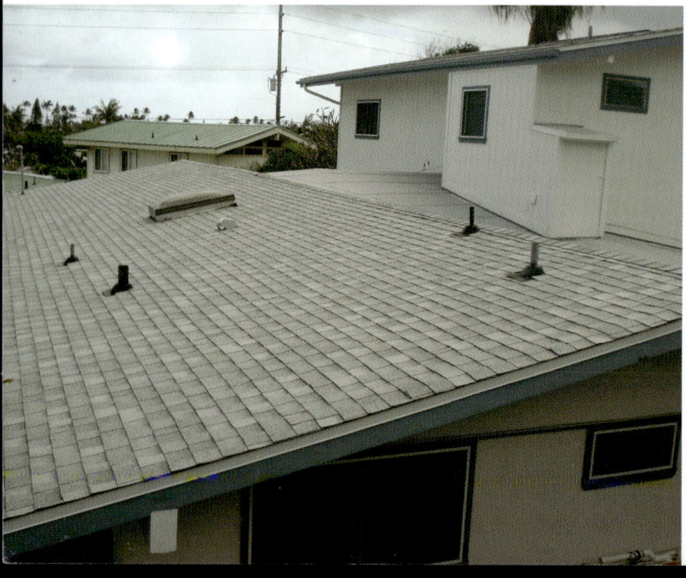
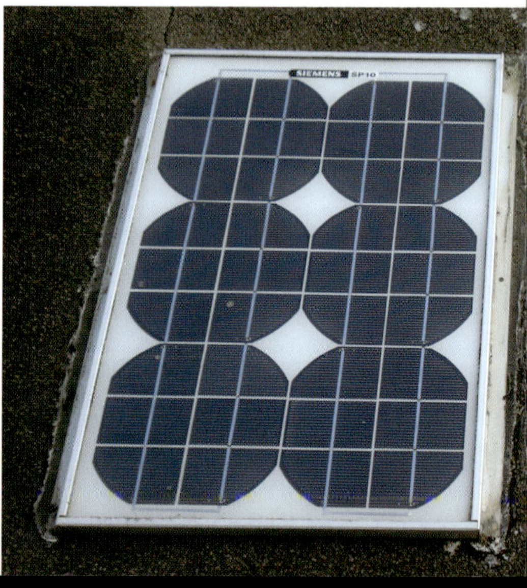
SIEMENS SP10

IN THE
HOME

part 1

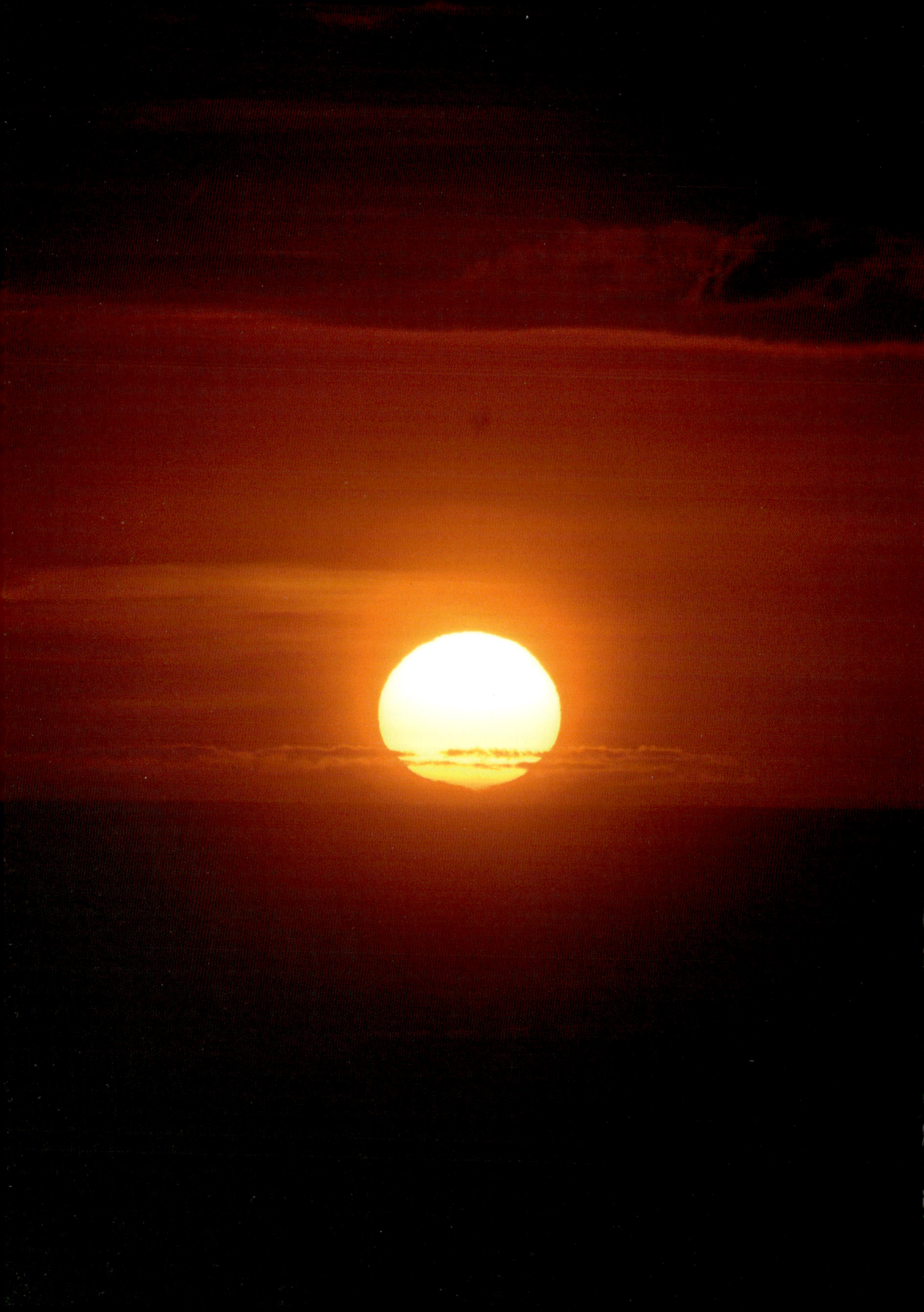

one: energy

Energy is ubiquitous. From the rhythmic electrical pulse of each heartbeat to the flick of a light switch, without energy in one form or another, life is not possible. The sun's energy has made it possible for life to flourish on this planet and for the human race to lead a comfortable existence. That same energy enters our bodies through the food we eat and leaves our body as heat.

Compact florescent light bulb.

Energy is a system. In the universe, energy is constant—it can neither be created nor destroyed. But with our blooming population and the consumption of energy that parallels our growth, it has recently become necessary to conserve resources from which we derive energy, like fossil fuels, and look to other viable options for powering our plugged-in modern day lives.

in the home • 4

Solar roof: solar attic fan and solar light tube.

Renewable energy is not a new concept, but it is becoming more widely accepted and practiced. People's awareness of the environmental impacts of producing energy derived from burning fossil fuels, our basic and most widely used system for energy production; the finite pool of resources; and the growing costs of living in an oil-dependent world have taken precedence over yesteryear's consume-without-prejudice mentality.

Hawai'i's energy situation is unique, as our state relies heavily on imported petroleum to meet our energy production needs. Without widespread access to fuel sources such as natural gas or large rivers for hydropower, we still have an opportunity to utilize the most powerful renewable energy source available—the sun.

By taking advantage of solar power with photovoltaic systems, solar hot-water heaters, solar attic fans, and solar light tubes, each family and every home has the opportunity to make Hawai'i a cleaner and more sustainable place to live. Cool, energy-efficient lighting with fluorescent light bulbs is a tremendously greener alternative to the standard incandescent lighting and one of the easiest ways to save electricity. Energy Star-qualified appliances conserve energy throughout the home and the utilities and state and federal governments offer rebates and tax credit incentives to make the conversion to a greener home easy on your checkbook.

Put time and care into your research and buying decisions. Realistically, the average family will take baby steps toward the goal of an energy efficient home. That's OK. Start by taking care of the appliances that give you the greatest return in reduced energy consumption and quick monetary rate of returns.

Ron Richmond, former executive director of the Hawaii Solar Energy Association, explains, "Conservation first; use what energy you want to without sacrificing quality of life, but don't waste. It's common sense. We are a first-world country; let's stay that way. Get the solar water heating system. On an annual average basis over the life of the system, there's a 90% contribution from the solar water heating system. Convert all your heater lights, a.k.a. incandescent lights, to compact fluorescents. And of course, convert to Energy Star appliances.

"Once you've done all those things, then take a look at solar electricity, but not before. If you have an energy bucket, you don't want to have any holes in that bucket before you start filling it up with solar stuff. Plug all the holes in your energy bucket first then start doing the PV stuff. That makes much more sense."

Think of your home as a bucket full of water, except instead of water, it's full of energy. A bucket with holes in it will leak and water will spill out—wasted. It's the same principle with the energy bucket. If you have an old second refrigerator in the hot garage, an electric water heater or a house full of incandescent lighting, in essence, you have holes in your energy bucket. Your home is wasting energy. Plug the holes in your energy bucket by switching to a solar hot water heater and recycling the inefficient second refrigerator, decreasing the amount of energy you consume and truly living in a "green" home.

There are many reasons for choosing to live in a sustainable manner. But it cannot be ignored that using renewable energy shows a deep respect for the land and our environment, and when that respect is magnified in the community, the collective difference is a change for the better: the conservation and preservation of life for generations to come.

Solar Hot Water Heater

According to the Hawaiian Electric Company, a conventional hot water heater is the biggest draw of electricity in the home. Up to 40% of the average monthly utility bill is due entirely to this appliance. And while hot water is not a necessity, most would be loath to go about their daily lives without it.

> Solar hot water systems save up to 80% to 90% of water heating costs. The energy savings can pay back the installation cost in less than two years.

> A utility-approved solar water heater is all that is needed for a non-air conditioned Hawai'i home to be considered an Energy Star home and to qualify for Energy Star mortgages.

in the home • 6

> To put things in context, Hawai'i has the largest number of solar water heating systems per-capita compared to any other state in the nation—close to 90,000 water-heating systems. Hawai'i has had a very positive experience with solar water heating, unlike other areas of the country.
>
> —Ron Richmond

PV panels.

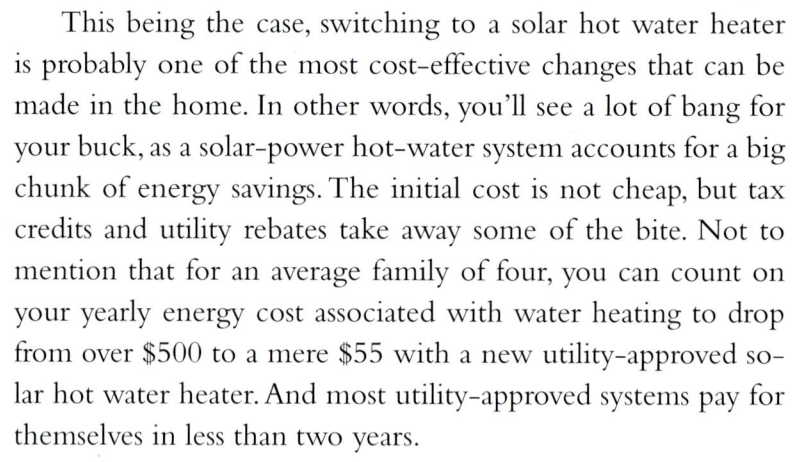

This being the case, switching to a solar hot water heater is probably one of the most cost-effective changes that can be made in the home. In other words, you'll see a lot of bang for your buck, as a solar-power hot-water system accounts for a big chunk of energy savings. The initial cost is not cheap, but tax credits and utility rebates take away some of the bite. Not to mention that for an average family of four, you can count on your yearly energy cost associated with water heating to drop from over $500 to a mere $55 with a new utility-approved solar hot water heater. And most utility-approved systems pay for themselves in less than two years.

"The estimated number of solar water heating systems installed in the U.S. in 2006 was 9,000, and 6,000 were from Hawai'i, so everyone across the country looks to Hawai'i as the standard," totes Ron Richmond, solar water heating and photovoltaic systems specialist since 1979. "The equipment has been extremely reliable, more reliable than the utility initially thought. All the equipment that was in place in the twentieth century is still appropriate technology for the first part of the twenty-first century. This is very proven, mature technology and extremely viable. The utility standard here in Hawai'i is the highest in the nation."

Hawai'i has been able to set the bar for solar water heating systems because of our consistent subtropical climate; namely, it doesn't freeze. Our rebate programs are also the best in the nation, so you can rest assured your investment in creating an energy efficient home is a sound one.

"The utility programs [in Hawai'i] have standardized the way the systems are designed and installed, and what equipment can be used. They do a 100-point system post-installation

7 • energy

inspection to ensure it's in compliance with their standards and specifications," Richmond continues.

The system works by heating up water in roof-mounted solar thermal collectors, a series of small tubes in a rectangular box with a transparent cover. The tubes are attached to an absorber plate, which is painted black to aid in heat absorption. As the water moves through the tubes, it is heated and stored in an insulated storage tank, ready for the house's hot water demands.

There are two basic types of systems: active and passive. A passive system is a thermal siphon system where the tank is located above the collectors on the roof. As the water is heated, it rises into the tank through convection and uses gravity to flow into the home. They are cheaper and require less maintenance, but they are also less efficient. Active systems, which are more common in Hawai'i, use a pump to move the water between the solar collector and the storage tank in a circulating loop. Storage tanks are often wrapped in a special insulating jacket to improve energy efficiency by reducing ambient heat loss.

First things first: determining the size of system you need to meet 100% of your family's hot water needs. Two factors come into play: the size of the solar collector and the size of the storage tank. The U.S. Department of Energy recommends around 20 square feet of collector area for each of the first two family members. For every additional person, add 8 square feet. A small (50- to 60-gallon) storage tank is usually sufficient for one to three people. A medium (80-gallon) storage tank works well for three to four people. A large tank is appropriate for four to six people. For active systems, the size of the solar storage tank increases with the size of the collector—typically 1.5 to 2 gallons of storage per square foot of collector.

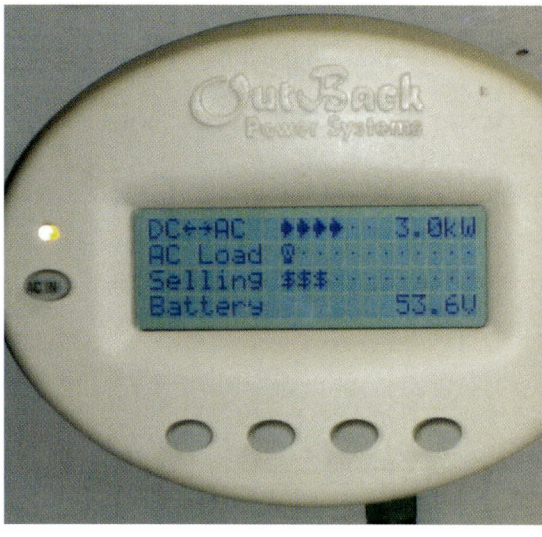

Power system display unit.

If you don't own a solar hot water heater, but are working toward energy efficiency, install a timer on the tank so that the water only heats up for one hour, once or twice a day, depending on the number of people in the household.

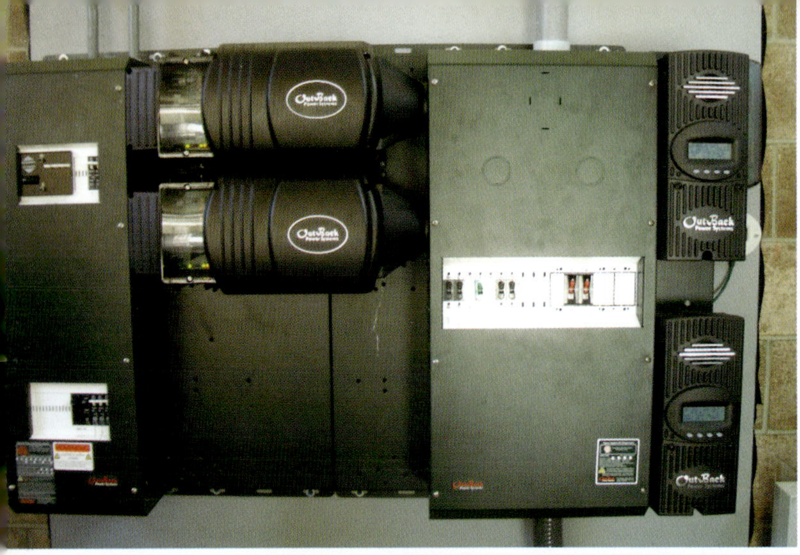
OutBack
OutBack
OutBack
OutBack

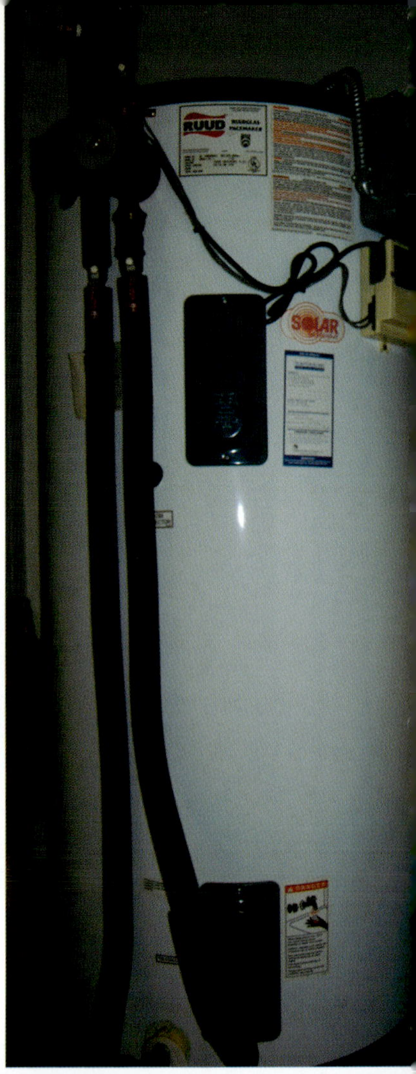
RUUD
SOLAR

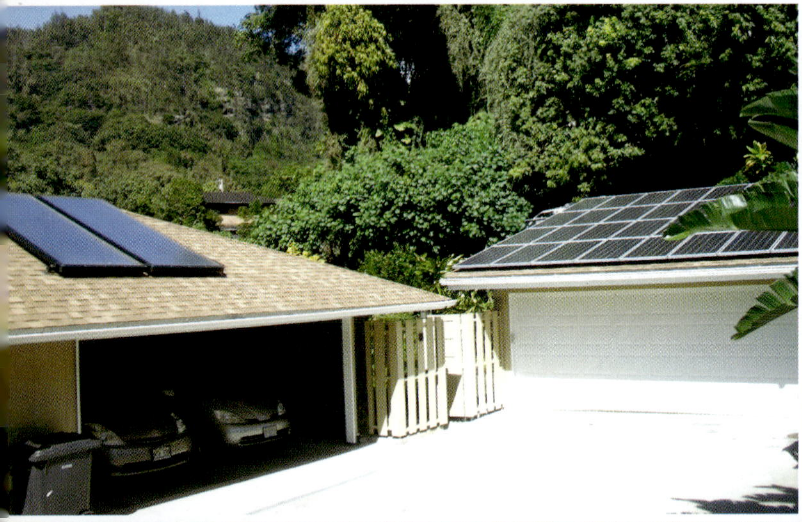

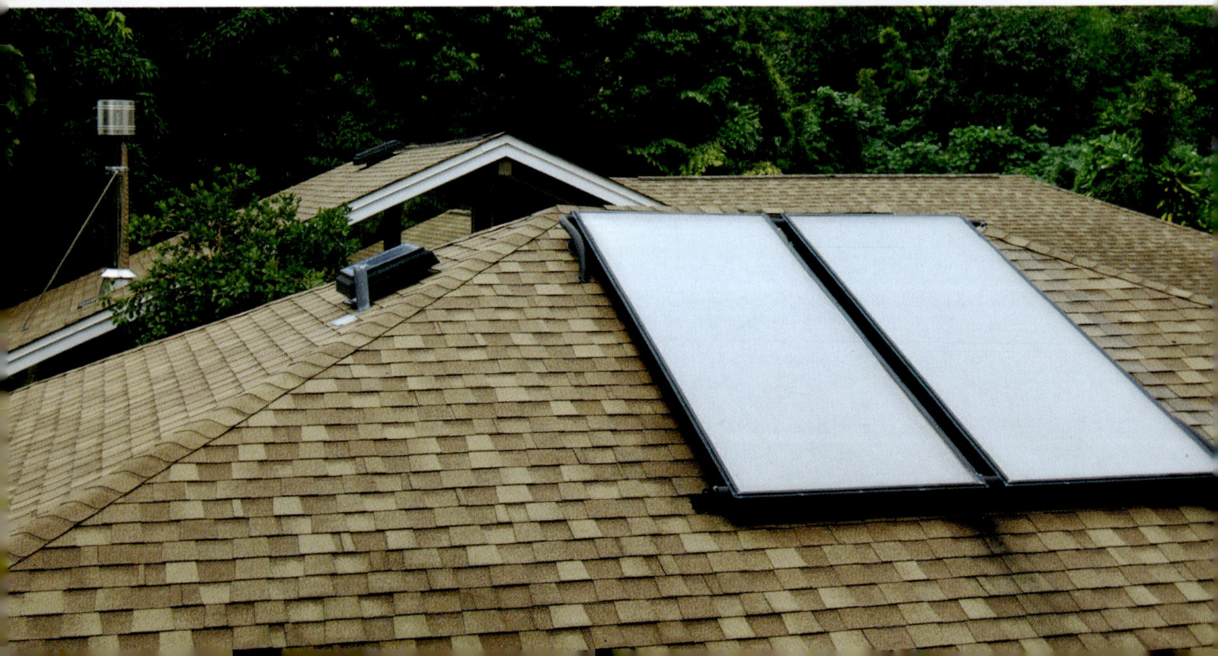

"It's easy and attainable," says Richmond. "It's easy because it only requires a one-day installation. Solar hot water heating is a home improvement. A lot of people think dealing with contractors can be time consuming, involving, and complicated. But with solar hot-water heating, once the customer decides to have it installed and the contractor schedules the installation, it's a one-day installation."

It is imperative to use a certified solar hot-water system contractor to install the system so it will work properly and operate efficiently. Be aware that some warranties and utility rebates become invalid if the system is installed by someone other than a certified contractor.

Solar Attic Fans

The attic of a house heats up throughout the day as the sun beats down on the roof. Radiant energy from the heat builds up in the attic space and can cause your living space inside the house to heat up and require additional energy consumption to cool. Solar attic fans help release some of that latent hot air using the same energy that created the heat in the first place—the sun.

There are many types of solar attic fans on the market and they are very reasonably priced for the work they perform. They require no electricity, as the solar panel creates all the power needed to turn the fan and circulate the attic air. No matter what style or model you choose, the purpose and function remain the same: to provide ventilation without increasing the energy load of the home.

"A solar-powered attic fan, especially a 20-watt, the biggest one available in the industry, will handle up to 1,700 square feet of living space, so if you live in a 1,700-square-foot house all under attic, one fan will take care of the entire house," explains Ray Heitzman, independent solar contractor and owner of Ray's Solar Fans. "Depending on the area you live in, the attic goes from 125 to 155 degrees. That radiated heat is like an oven on top

---

HECO offers a $3 credit every month for those eligible and willing to install a free Energy Scout radio device that can temporarily turn off your water heater during times of emergency.

Insulating your water heater's storage tank can reduce stand-by heat loss by 25% to 45%, which translates to 4% to 9% in heating cost savings. If your water heater's storage tank is warm to the touch, it needs insulation.

Hawaiian Electric Company (HECO) offers a $1,000 rebate when you use a HECO-approved participating contractor to install the unit. Combined with a 30% federal tax credit and a 35% Hawai'i state energy tax credit, it turns a $5,000 expense into a $1,500 expense. Not bad!!

in the home • 10

Solar attic fan.

A $900, 20-watt solar attic fan drops to less than $400 after a 30% federal tax credit and a 35% Hawai'i state tax credit.

of your head, radiating down right through the drywall and into your living area. When you get an attic fan it's like popping a champagne cork because you have an immediate cooling effect. Within 15 minutes you'll feel a difference—it's amazing."

Solar attic fans have a small solar panel that captures the sunlight from dawn to dusk and converts it to energy. The bonus with solar attic fans is that they always work the hardest when necessary. On sunny days, when the attic space heats up quickly from the radiant heat of the sun, the solar-powered attic fans turn their fastest to move and release that trapped, stale air. When the clouds make a presence and the attic stays relatively cool, the fans don't need to work so hard, and they don't.

Laurie Carlson, publisher of the *Honolulu Weekly* and *Hawai'i Island Journal*, recently installed two solar attic fans in her home and relishes the cool comfort they provide. "Before, when you would come into the house on a hot afternoon, you could really feel the heat above your head. And now, having solar fans to let the air out of the attic, we don't have that oppressive blanket of air above us—it's really quite comfortable. It's made such a difference in our living space; it's well worth it what ever the cost may be."

Attic ventilation is not only important for removing the unwanted heat to keep the attic cool, which in turn keeps the house cooler; but the added circulation also prevents mold and mildew from growing in damp locales. Solar attic fans can be wind resistant, a good choice for windward residents who catch the full brunt of the trades, as well as impact resistant for the odd falling coconut.

Solar attic fans are rated by the amount of air (in cubic feet) they move per minute. Most will circulate at least 800 cubic feet of air a minute. Determine the cubic feet of your attic space by multiplying length, width, and height, then purchase enough solar fans to properly circulate, ventilate, and cool the attic space. Make sure to distribute them evenly for proper circulation and mount them high on the roof, near the ridge, for best results.

"When you install a solar attic fan, you're doing a number of things," says Heitzman. "You're taking all the hot air out of the attic, and you're taking all the moisture out so you're killing any black mold growing in your attic, which happens all the time. You are also preserving your roof, with 30% more longevity because your roof dies from the inside out; it gets baked from the inside out. You are also reducing the compressors on your air conditioning unit by 30% to 40% because they don't have to fight against the oven that was once radiating on top of your roof."

Most solar attic fans are simple enough to install yourself. But if you feel uncomfortable on the roof or in the attic, always use a licensed contractor to do the job. Besides, the state of Hawai'i offers tax incentives on solar attic fans installed by a qualified solar contractor.

Solar-powered attic fans work best in conjunction with other attic vents under the eaves or the roof, soffit vents, or gable vents that draw in cooler air blowing by. A natural convective process releases the warmer air out of the solar vents to cool the attic. That will be covered in further detail in the design chapter.

> Heat radiates very well. It radiates right through, just like a hot plate radiates heat into a pot and boils water. The same is true of heat radiating through your attic into your home. Once you relive that pressure and remove that heat, the attic temperature will drop 30, 40, even 50 degrees, which results in a 7- to 10-degree minimum drop in your living area. And it also means 30% to 40% less strain on the air conditioner because the compressors can turn on and off more instead of staying on all the time.
>
> —Ray Heitzman

Electric Lighting
Compact Fluorescent Light (CFL) Bulbs
One of the easiest things you can do at home that will make an incredible difference in reducing your electric consumption and utility bill is simply replacing your conventional incandescent light bulbs with energy-efficient compact fluorescent light (CFL) bulbs. CFL bulbs use about one quarter of the energy that it takes to light up a standard incandescent bulb, and they can last up to ten times longer. Because of the method in which compact fluorescent bulbs produce light, they are much cooler

in the home • 12

than their incandescent counterparts. In our warm Hawaiian climate, this reduces the cost of cooling your home.

Ron Richmond is a proponent of CFL bulbs. "CFLs reduce your electric consumption without sacrificing the level or quality of light. They also reduce the amount of heat that is generated in the house by incandescent lighting, because incandescent bulbs are primarily heaters and light is just a by-product."

CFLs come in all shapes and sizes and can be used in just about any fixture. You are probably familiar with the basic mini-spiral shape, the poster child of the CFL boom, but that is just the tip of the iceberg. These fluorescent bulbs are also available in a small vertical tube, a mimic of a standard incandescent bulb, a globe, and a candelabrum. For recessed lighting, a reflector CFL is suggested instead of a spiral so that the bulb evenly distributes the light to the task area. Reflector bulbs are also made for outdoor use. For three-way switches and light fixtures connected to a dimmer, make sure that you select CFL bulbs that specify use with three-way switches and dimmers. Regular CFLs are not designed to work with these types of switches and will burn out.

Some people assume that because CFL bulbs are indeed a type of fluorescent light, the only shade of light they emit is the bluish-white hue found in office buildings and warehouses, a shade of light not very relaxing or comfortable for the home. But CFL technology has come along way and now CFLs are available in soft or warm white light, the shade of light similar to the standard incandescent bulb that we are so familiar with. Because different colors in the home are enhanced by different shades of light, CFLs are also available in the following brighter shades: bright white, natural, and daylight.

The light output of a CFL bulb is given a number called a lumen rating, which correlates with wattage of incandescent bulbs. The higher the lumen rating, the higher the light output. When replacing incandescent bulbs with CFL bulbs, you'll need to find the right lumen rating that is equivalent to the old bulb. This information is usually included by the manufac-

---

HECO offers coupons for Energy Star CFL bulbs on their website, www.heco.com, to help you save a few extra bucks at the register, in addition to the long-term and environmental savings.

In Australia, the government took action and, in an effort to convert the country to CFL light bulbs, they went house-to-house and changed every light bulb across Australia—for free!

turer on the package using wordage like "60 watt replacement." Lumen rating and wattage tables are available on the Energy Star website, www.energystar.gov.

CFL bulbs do contain a trace amount of mercury, though much less than that in a household thermometer, and they need to be disposed of properly. At the time of publishing this book, the State of Hawai'i does not offer CFL bulb recycling. In the meantime, the EPA suggests several ways to safely dispose of used bulbs: place used bulbs in two plastic bags, seal the bags and then dispose of in the trash; or seek out a fee-based CFL recycling program. Recycleyourcfl.com offers a CFL bulb recycling kit that can hold up to fifteen small bulbs or six to eight medium to large bulbs. All you have to do is pack up your used bulbs and once the box is full, ship it off.

If a CFL breaks in your home, open a window and ventilate the room for at least 15 minutes. Turn off the air conditioner if you have one. Scoop up—don't vacuum—the broken glass onto stiff paper or cardboard and place the fragments in a glass jar with a lid or seal in a plastic bag, including the cardboard. Use packing tape or duct tape to pick up the tiny pieces and dust that didn't make it onto the paper. Seal the tape in a plastic bag and throw it away as well. If the bulb breaks over carpet and a vacuum must be used to get all the dust and small pieces, seal the used vacuum bag in a plastic bag and throw it away immediately after use. Take the rubbish outside and dump it in the can, then wash your hands.

A handful of CFLs.

Install motion sensors on outdoor lights and cut the use of a 150-watt, outdoor flood light from six hours to one hour per night. You will save 270 kilowatt/hours and $38 per year.

-Courtesy of HECO

Halogen Bulbs

If you can't find CFL bulbs designed specifically for dimmer or three-way switches, consider halogen bulbs as an alternative. Halogen light bulbs are a type of incandescent light bulb filled with halogen gas, which preserves the filament, resulting in light bulbs that last two to three times longer than regular incandescent bulbs. They are also 10 to 20 percent more energy efficient than their incandescent cousins.

Halogen bulbs work well in track and task lighting where dimmer switches control the amount of light needed for a specific task. Task lighting can even be wired so that the output of each individual light can be controlled separately from others on the same switch. When you can put a specific amount of light where you want it, then you are truly being energy efficient.

Energy Star Appliances

What exactly is Energy Star? Is it an appliance manufacturer? Maybe a general contracting company specializing in building energy-efficient homes? What about a brand of light bulbs? Just a cute logo? Wait, wait, the newest club in Las Vegas?

The Energy Star logo is popping up more and more in Hawai'i and around the country. It appears on refrigerators, washers, computers, windows, and even entire houses.

So, what really is Energy Star? It is a joint program between the U.S. Environmental Protection Agency (EPA) and the U.S. Department of Energy, with the mission of protecting the environment by curtailing energy production and consumption while saving consumers money and helping grow the economy with energy-efficient products and practices. When products like refrigerators, computers, and even entire houses meet the strict energy efficiency guidelines set forth through the government program, they qualify for the Energy Star branding and logo and become one of the best choices—financially and environmentally—for turning your home into a green machine.

If the old appliances in your home are numbered in their years, they are most likely big drains on your energy consumption and wallet, especially that second rusty refrigerator that's in the hot garage of all places. Replace them with Energy Star-qualified appliances that perform as well as or better than standard brands and save you money while also saving the environment. It

Energy Star branding.

15 • energy

just makes sense. Even if you have newer appliances, switching to Energy Star appliances will save you money in the long run by cutting your utility bill by about 30%. And don't forget to recycle your old appliances instead of taking them to the dump.

Listed on the following pages are some of major appliances we use in Hawai'i.

Ceiling Fans

The sub-tropical climate of the Hawaiian Islands is revered for just the right mix of heat and humidity, a comfortable combination that is one of our trademarks. But the Islands' unique geography creates microclimates, which can be drastically different from one another. There are cold and windy mountainous regions as well as hot and dry leeward coastal plains. We have rainforests and we have lowland scrub forests.

Ceiling fan.

No matter what the locale, when the sun shines down on our Islands and our homes, it's intense, and our houses absorb that heat energy. But instead of cooling the house down with air conditioning, an energy-sucking appliance, consider an energy-efficient ceiling fan in every room.

Ceiling fans draw hardly any energy at all, and now Energy Star offers ceiling fans that are 50% more efficient than standard ceiling fans. For our tropical climate, use the ceiling fan in the counter-clockwise direction. While standing directly under the ceiling fan you should feel a cool breeze. The airflow produced creates a wind-chill effect, making you "feel" cooler.

The fan blades are what classify the size of the ceiling fan and are designed between 29 and 54 inches long. The length of blade should correspond to the size of the room where the fan will go. Ceiling fans should be installed or mounted in the middle of the room and at least seven feet above the floor and 18 inches from the walls. If the ceiling height allows, install the fan eight to nine feet above the floor and eight to ten inches below

HECO offers a $40 rebate on the purchase of a new Energy Star ceiling fan. Check the www.energystar.gov website to be sure the brand and model you plan to purchase is listed, and then go to www.heco.com for the rebate application.

the ceiling for optimal airflow. Proper mounting and wiring are critical for maximum energy efficiency.

For best results at beating the heat without an air conditioner, ceiling fans should be used with other methods of cooling the home, such as radiant barriers, solar-powered attic fans, and window coverings.

Refrigerators

In most homes, this appliance is the biggest draw on energy in the kitchen. Energy Star-qualified refrigerators have high-efficiency compressors, improved insulation, and more precise temperature and defrost mechanisms, which results in 15% less energy used than required by current federal standards, and 40% less energy than conventional models sold in 2001. Energy Star also puts its mark on freezers and compact refrigerator/freezer models that perform more efficiently than federal standards mandate.

> Earn a $50 rebate from HECO when you purchase a new Energy Star-qualified clothes washer. Check the www.energystar.gov website to be sure the brand and model you plan to purchase is listed, and then go to www.heco.com for the rebate application.

No matter what model of refrigerator or freezer you own, here are a few ways to improve its energy efficiency:

1) Position your refrigerator away from a heat source such as an oven, dishwasher, or direct sunlight from a window.

2) Leave a space between the wall or cabinets and the refrigerator or freezer to allow air to circulate around the condenser coils.

3) Make sure the door seals are airtight and clean.

4) Keep the refrigerator's thermostat set between 35 and 38 degrees Fahrenheit and the freezer at 0 degrees Fahrenheit.

5) Minimize the amount of time the refrigerator door is open.

6) Always recycle older or second refrigerators.

—Courtesy of the EPA

Clothes Washers

Energy Star clothes washers use 40% less energy than standard washers and use less water per load. Full-sized Energy Star-qualified washers use 18 to 25 gallons per load where standard washers use 40 gallons, an added bonus to the energy saved. They extract more water from clothes during the spin cycle, reducing drying time and wear-and-tear on your cool threads. Washing with full loads is the most energy-efficient way to do your washing, saving water, energy, and time. Using cold water as often as possible saves energy as well.

> Energy Star does not label clothes dryers, but here are some ways to reduce the amount of energy it takes to dry your clothes:
>
> 1) Use the moisture sensor option on your dryer, which automatically shuts off the dryer when your clothes are dry.
>
> 2) If your washer has the option of a high-speed spin cycle or extended spin cycle, use that feature to reduce the amount of remaining moisture and speed drying time.
>
> 3) Air-dry whenever possible.
>
> —Courtesy of the EPA

Dishwashers

The main draw of energy from dishwashers comes from heating the water used to wash the dishes. Energy Star dishwashers are 41% more energy efficient than the federal minimum standards for energy consumption for conventional dishwashers because they use less hot water to get the job done. In addition, they use less water per cycle, which saves water as well as energy.

For maximum energy efficiency and water conservation, always run your dishwasher with a full load, and choose the appliance's air-dry option instead of heat-drying.

Room Air Conditioners

In a perfect Hawai'i, everyone would do away with their cumbersome air conditioning units and utilize the cooling effect of ceiling fans and solar attic fans instead. But obviously, this is not

HECO offers a $50 rebate on the purchase of an Energy Star-qualified dishwasher. Check the www.energystar.gov website to be sure the brand and model you plan to purchase is listed, and then go to www.heco.com for the rebate application.

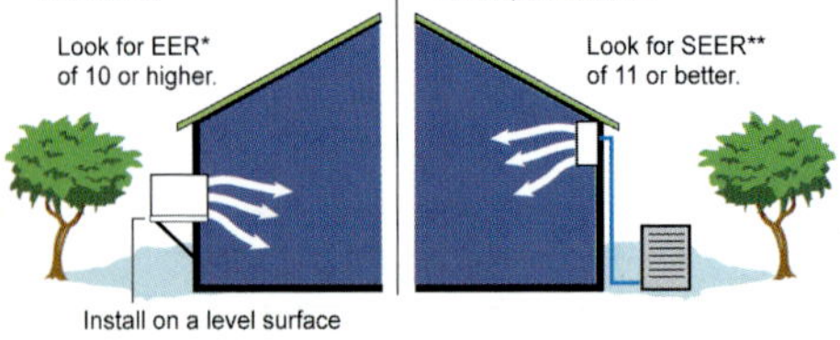

Window Air Conditioner
Performs most efficiently in a shaded location, like on the north side of a home.

Look for EER* of 10 or higher.

Install on a level surface

*Energy Efficiency Ratio

Ductless Split System AC
Consider units with multi-speed fans & compressors for better overall performance.

Look for SEER** of 11 or better.

**Seasonal Energy Efficiency Ratio

How much AC do I need?

floor area (in square feet)	Capacity (Btu/hr.)
150 - 250	5,000
200 - 300	6,000
250 - 350	7,000
300 - 400	8,000
350 - 450	9,000
400 - 550	10,000
500 - 650	12,000
575 - 800	14,000
750 - 1000	18,000

the case, and some people would be hard pressed to part ways with their air conditioning unit.

For those homes that find air conditioning is a must, there is still a way to cool your home with Energy Star-qualified air conditioning systems and be energy efficient while doing it.

Energy Star-qualified air conditioners use 10% less energy than conventional units by using timers for better temperature control, utilizing the minimum amount of energy needed to cool a particular room.

The key to an energy-efficient air conditioner is using the right size unit for the square footage of the room to be cooled. First, determine the square footage by multiplying the length by the width of the area to be cooled. For a triangular room, multiply the length by the width and then divide by two. Next, use the chart on the left to select a unit with the corresponding cooling capacity. The cooling capacity for air conditioners is measured in BTUs (British Thermal Units).

Some people think a larger unit will cool a room better, but this is not the case. Air conditioners cool the air by removing heat and humidity at the same time. Too large of a unit cools the air too quickly, leaving the humidity behind and a damp and clammy feeling in the room. It uses a lot of energy to do an ineffective job.

Here are some guidelines to account for the little nuances in life:

1) If the room is heavily shaded, reduce the cooling capacity by 10%.

2) If the room is very sunny, increase the cooling capacity by 10%.

3) If more than two people occupy the room, add 660 BTUs for each person.

4) If the unit is used in the kitchen, increase the capacity by 4,000 BTUs.

5) Take into account where in the room the unit is being installed, and direct airflow in the proper direction.

—Courtesy of the EPA

If possible, install your window air conditioning unit in a shady location for maximum efficiency. Trees and shrubs are a great way to obtain shade and beautify your home. Just watch out that leaves and debris don't clog the air intake.

Computers

Whether you're a fan of the PC or Mac, computers have become a necessity in the home for education and entertainment. They have become as common as cars or cell phones, modern conveniences it seems we cannot live without.

And most likely because of its small size and often inconspicuous and sleek design, computers are often overlooked as an appliance. But in fact they are plugged in and are a continual draw on power.

Adding fuel to the fire of wasteful energy practices, some people contend that it is better to leave the computer turned on and running day and night, as turning it on and off could damage the hard drive. The debate is a lively one; all you have to do is check a few blogs to get a healthy dose of both sides of the argument.

In 1992, computers and monitors were the first products to receive the Energy Star label. Since then, the EPA has conducted research on computer usage trends and found that per-day usage has soared, along with the demand for energy to power the increase. To combat the heavy usage, Energy Star implenented even more stringent energy efficiency requirements for computers and laptops beginning in July 2007.

Today's Energy Star-qualified computers must meet energy usage guidelines in three operating modes: standby, sleep mode, and while the computers are in use. They must also include a more efficient internal power supply. These new computer specifications are estimated to save consumers and businesses more than $1.8 billion in energy costs over the next five years and prevent greenhouse gas emmisions equal to the annual emmisions of 2.7 million vehicles.

> Purchase a new Energy Star-qualified window air conditioner and save $75 with a rebate from HECO. Check the www.energystar.gov website to be sure the brand and model you plan to purchase is listed, and then go to www.heco.com for the rebate application.

> About 12% of the average home's electricity costs involve air conditioning.
>
> —Courtesy of HECO

Other Useful Appliances

In Hawai'i we are obviously not in the market for heating systems or snow blowers, so here is a list of other Energy Star appliances available to us that you might not be aware of:

- Dehumidifiers
- Battery charging systems for power tools and small home appliances
- Central air conditioning systems
- Ventilating fans used in the kitchen and bathroom
- Programmable thermostats
- Light fixtures
- Home electronics like cordless phones, DVD players, and televisions
- Home office equipment like copiers, fax machines, monitors, printers, and scanners
- Water coolers

Photovoltaic Systems (PV)

"Electricity is created by burning fossil fuel and taking the heat energy of boiling water, turning a turbine, and then cooling the water down and recycling it through a typical carbon cycle," explains John Harrison, conservationist and environmental coordinator at the University of Hawai'i at Mānoa's Environmental Center. "It's inherently inefficient. The general quoted efficiency for most conventional power plants like the ones that HECO runs is in the area of 33% to 36%, and that means that two-thirds of the energy goes out the smoke stack right at the start. It's not the best use of a fossil fuel."

Harrison is a proponent of taking advantage of Hawai'i's most prevalent natural resource, the sun. He is talking the talk and walking the walk, having converted his home to a self-sufficient photovoltaic system and using his completely "green" abode to teach others about the benefits of solar power and energy neutrality.

"I have 32 panels, which is a five kilowatt system. The system is divided into two subsystems and each one is two-and-a-

---

In 2006, Americans who used Energy Star products saved enough energy to prevent 37 million metric tons of greenhouse gas emmissions equivalent to those from 25 million cars—saving a total of $14 billion on utility bills. They also saved more than 170 billion kilowatt hours, which is almost 5% of the total 2006 electricity demand.

The typical household spends about $1,900 a year on energy bills. With Energy Star appliances, you can save up to 30%, or more than $600 a year.

—Courtesy of the EPA

half kilowatts. It's all off-the-shelf stuff, there is no custom design on any of it. They are all conventionally and commercially available," says Harrison.

What does it take to put together a photovoltaic system? A basic list includes an array of photovoltaic panels, an inverter, a distribution panel, and a computer panel to monitor the system. Better yet, add batteries, a battery charger, and a switching box to the list.

Here's how it works. Photovoltaic cells are grouped together into panels. The entire group of panels used in a system is called an array. Arrays will vary in size depending on the desired power output of the system, which is based on the energy consumption of a particular household. The array is positioned on the roof of a structure and oriented to the south to capture the greatest amount of sunlight.

Photovoltaic cells capture sunlight and convert it to direct current (DC) electricity. In the U.S. we use alternating current (AC) electricity. The DC electricity is sent to a converter where it is transferred to AC and made ready for use. A computer distribution panel sends the electricity to the meter before it is available for use in the home.

For photovoltaic systems with battery backups, electricity that comes from the panels first goes through a switching box controlled by a computer, which senses the float level (maximum charge) of the batteries. If the batteries need to be charged, the computer diverts enough of the incoming power during the daylight hours through the battery chargers to bring the batteries back up to maximum charge, and simultaneously sends power to the inverter to meet the power demands of the house. If the batteries are already at float level, the switching box sends all the power through the inverters for distribution to the house. Battery backups come in handy on rainy days or during power outages, when the PV system cannot produce enough energy to meet the demands of the household.

On a typical sunny day, the electricity produced by the system would charge the batteries to maximum capacity and be

---

The State of Hawai'i, Department of Business, Economic Development and Tourism has a great website for all sorts of links concerning the facets of using solar as renewable energy: Hawai'i state and federal tax credit information, approved solar water heating contractors on all islands, a great solar map showing solar radiation for each island, a complete list of on-line solar publications and links to several other solar sites. Make sure and give this a thorough look: www.hawaii.gov/dbedt/info/energy/renewable/solar/.

in the home • 22

Having a PV system installed in your home is a big investment and should only be undertaken after you've implemented other paramount design and energy-efficient principles throughout your home. The cost of a PV system is determined by many factors in and around the home and the size of the system you need to power your life. Generally you can expect to pay $8 per watt for a system without a battery backup and $10 per watt for a system with a battery backup. That's $8,000 to $10,000 for a one-kilowatt system. Keep in mind that John Harrison powered his home with a five-kilowatt system.

available for the needs of the home. If the residence is still connected to the grid, any electricity produced above and beyond the needs of the home is fed back into the main grid and made available for use by other traditional electric utility customers. In this way, the individual becomes a producer of electricity.

"There are two different ways you can have connection to an existing utility under Hawai'i law," explains Harrison. "One is called net metering, which is what we do. If we produce more than we consume, then we don't pay, except we do pay the standard connection fee. If we consume more than we use, during bad weather or winter time, then we pay the normal residential rate, which is about 23 cents a kilowatt/hour. The other way is a purchase power agreement. In that system they meter what you produce and what you consume individually. They pay you for what you produce and feed into the grid, and you pay them for what you consume and take off the grid. The kicker there is they pay you the wholesale rate, which is about 11 cents a kilowatt/hour. You pay them at the retail rate which is about 23 cents a kilowatt/hour, which means to break even you have to produce twice as much energy as you consume."

The energy measured in kilowatt/hours that an individual residence puts back into the grid is accumulated over the course of a year. This allows the residence to use electricity from the grid without paying for it as long as the residence has produced a surplus of energy. In other words, the residence doesn't pay until there is a debit on the individual system.

Harrison has designed a home that is an energy-efficient system in itself, with many sustainable facets of design and efficiency to create a place of energy neutrality. In fact, he is giving back to the grid more energy than he is consuming and has essentially created a self-sufficient and self-sustaining home. With no real need for the grid and its inefficient power source, Harrison still stays connected. "I could flip a switch and disconnect from the grid and I'd be fine, but the reason I don't do that

05
117697

HECO 518520
47 480 702
Use for Net Metering
NM
Itron SENTINEL® Multimeasurement Meter
CL 200 120-480V 3W TA 30 Kh 7.2
METER TYPE SS1S1L FM 2S 60 Hz
WARNING!
TAMPERING WITH HECO PROPERTY IS A CRIMINAL OFFENSE. YOU MAY BE PROSECUTED UNDER HAWAII REVISED STATUTES. HECO MAY REQUIRE YOU TO PAY TRIPLE DAMAGES. HRS 269-71

is very simple: if I did, all the extra energy I produce would be wasted. This way it goes back into the grid."

In June of 2001, Hawai'i became the 35th state to adopt a Net Energy Metering Law allowing renewable energy grid-connected systems up to 50 kilowatts. The average household might use 23 kilowatt/hours per day, and an energy-neutral home only 12 kilowatt/hours per day.

Make sure to use a licensed and fully insured solar energy contractor to install your PV system. Check your phone book for local listings.

Natural Gas

As far as land masses go, Hawai'i is one of the youngest, obviously apparent in the lava that continues to ooze forth from Kīlauea and the submarine volcano Lō'ihi, some 15 miles southeast of Hawai'i and more than 3,000 feet below the ocean's surface, slowly expanding the Big Island and the Hawaiian archipelago. Because of our volcanic origins in the middle of the Pacific Ocean, natural gas is not a naturally occuring phenomenon, nor is it a common source of energy, as it on the mainland. But the Gas Company has found a way to supply the greater Honolulu area with synthetic natural gas (SNG).

The Gas Company produces synthetic natural gas, a product of refined crude oil, at its manufacturing plant on O'ahu. Honolulu residential utility customers are delivered their synthetic natural gas via an underground pipeline system. Those not connected to the pipeline system are delivered propane in cylinders or tanks. The neighboring islands are delivered propane through an underground pipeline as well.

While nearly 90% of the demand for natural gas comes from the industrial and commercial sectors (think Waikīkī) of Hawai'i, it is also a viable option in the home for running several high energy-consuming appliances.

The first appliances that will most likely come to mind are gas cooking appliances in the kitchen. Gas ranges offer preci-

> A barrel of oil converted to gas goes a lot farther than a barrel of oil converted to electricity. In other words, gas will take up to 40% less oil than electricity to do the same work.
>
> —Courtesy of the Gas Company

sion performance through complete control of the flame, and still work during a power outage. It's no wonder that most restaurants, with their high demand for quality and quantity, use gas cooking equipment.

Laurie Carlson explains why she switched over to a gas range in her kitchen: "The thing about gas and propane that is so great is that when you burn oil to make electricity, you lose a lot of energy in the transfer of energy from one way to another. You're burning oil to make electricity and you're putting it in a line, bringing it into the house, and you're turning it on to make heat on an electric range and the energy conversion loss is huge. With gas or propane, you're taking the gas and using it right there, you're not losing it. It's much more immediate and direct."

Being a resident on the windward side, Carlson's home is not connected to any natural gas pipeline, a predicament she easily sidestepped. "We bought a tank, not a huge tank, about five feet tall. It's small enough to throw into our van and take it to get refilled."

If you prefer, the Gas Company offers a service to deliver propane directly to your home, with a nominal tank rental fee.

Gas hot water heaters can save a lot of money over electrically powered units, and the savings go even further when the other energy-savings techniques are employed. For instance, turning down the thermostat from "hot" to "warm" (120 degress Fahrenheit) can reduce gas consumption by up to 8%; turning the thermostat down as far as it will go during non-use periods can reduce gas consumption up to 10%; insulating your water heater tank and pipes can reduce heat loss and cut energy use.

There are several other common appliances that will save energy and money when converted to gas. Gas clothes dryers will definitely cut down the energy consumption over electricity, and some newer gas dryers even feature pilotless ignition systems that can reduce gas energy usage by up to 15%. Pool and spa heaters, as well as outdoor barbeques, can also be run on gas.

> In the production process, odorant is added to the odorless natural gas to give it that "gas" smell for safety reasons.

> Carbon dioxide, a by-product of synthetic natural gas production, is sold to companies that manufacture dry ice and liquid carbon dioxide.

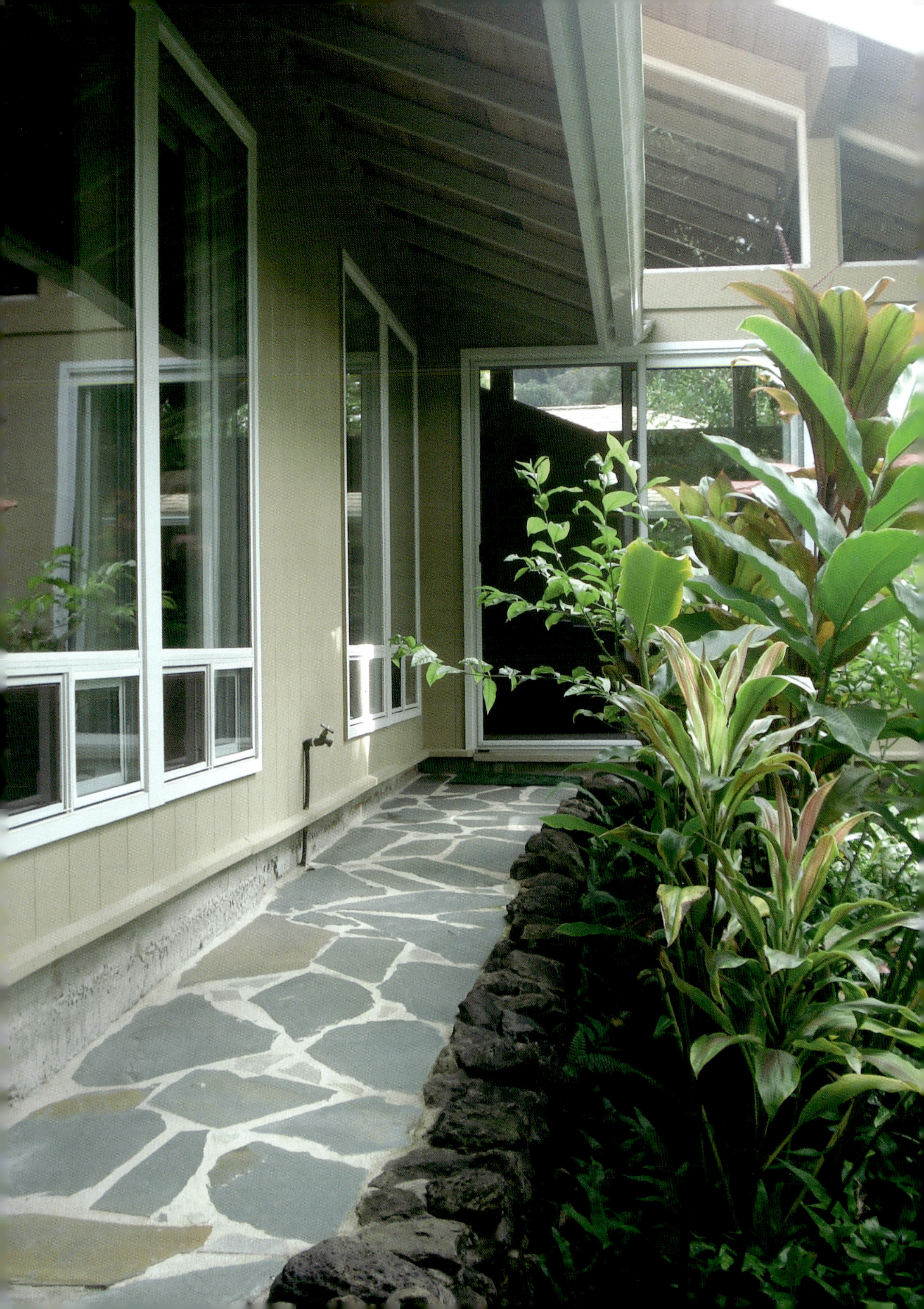

two: design

Energy efficiency in the home is not limited to energy-saving appliances and solar power. There are many passive systems that can work to cool and light living spaces naturally, without the need to plug in a cord and pull from the grid. Some of these design principles, like passive cooling through window and vent placement, might already be part of your home. Quite possibly, your home could be set up as a thermal chimney and you've yet to take advantage of the design. Skylights and solar light tubes, which bring diffused natural light into the home, can be installed at modest prices and brighten a home not only in the day, but also at night, taking the place of an electrical light.

Roofs with solar panels.

> LP TechShield is a common choice among "green" builders for its environmentally smart panel made of wood strands and resins and its ability to allow moisture to escape. It also carries the Energy Star brand.

As much as the sun can be a useful tool to power entire electrical systems in a home with photovoltaic technology, its radiant heat energy can be a double-edged sword, causing the temperatures in the attic and the house to rise dramatically throughout the day, damaging roofs and attics and creating an uncomfortable living space. As we harness the sun's energy to heat our water and power our appliances, we must also block the radiant heat from entering our homes. There are several ways to accomplish this goal.

Through specific design priciples, proper natural lighting, and attic ventilation, it is possible to cool a home using the cool rush of the trade winds, solar-powered attic fans, and radiant barriers instead of the electrical hum of the air conditioner.

John Harrison, who understands the idea of the solar bucket and recognizes the need to plug all the holes, employed many of the passive cooling and lighting techniques in the design of his green home. "The single most important part of the whole project is the recognition early-on of a total design concept. You can approach an energy remodel from a couple of different ways. One is as an add-on and that is going to have limitations because what is entailed requires a systems approach, which is much more comprehensive than just slapping on some panels and water heater."

Radiant Barrier

There are no ifs, and, or buts about it: direct sunlight in the Hawaiian Islands is hot and oppressive. Try this little sunlight experiment to see for yourself. Stand in the sunlight at 8 AM and notice the cool air around you and the comfortable warming feeling of the earth's surface heating up in the soft morning light. Now go to that same place at 2 PM and feel the difference. The sun has been battering that very spot all day. The earth has warmed significantly, and the air surrounding you is most likely at least ten degrees Fahrenheit warmer. Plants

Radiant barrier.

have closed their stomata, respiratory pores on the underside of leaves, to prevent evaporation. The overhead sun is hot and sharp and you begin to perspire within minutes.

What does this little experiment tell us? One, a quick dip in the ocean would be nice, and two, our homes are heat sinks, soaking up all those invasive sun rays and transferring heat into our homes. This daily occurrence of our homes heating up with the overhead march of the sun is the reason for all the energy we use, in electrical appliances like air conditioners and ceiling fans, to cool our homes.

It's time to mount a concerted effort against radiant heat and take back our living spaces to make them cool and comfortable with energy efficiency. One of the first lines of defense against radiant heat is a radiant barrier that keeps the sun's heat from entering the attic. When the attic is cooler, the entire house is cooler and all your appliances work less to get the job done, saving money and energy in the process.

John Harrison explains his use of Techshield, a-top-of-the-line radiant barrier, as it works together with other attic vents to create a passive cooling system. "The radiant transmission of energy through the roof is a big, big problem. In general, radiant seepage of insulation into the interior of the house is a problem. What we did was on top of the plywood subroof, laid parallel to the roofline, ¾-inch batons and used a material called Techshield, which is a recycled wood product that is specifically designed as a thermal barrier. It creates a ¾-inch space between the subroof and the Techshield. It has a reflective interior aluminum surface. The incoming heat radiant

The reflective side of the panel faces IN to the home on the roof. On a wall, the reflective side faces OUT.

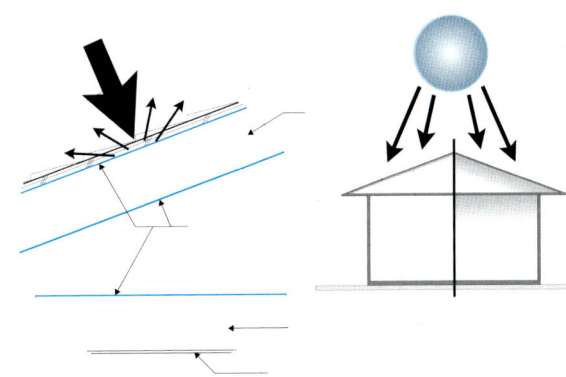

Roof radiant barrier.

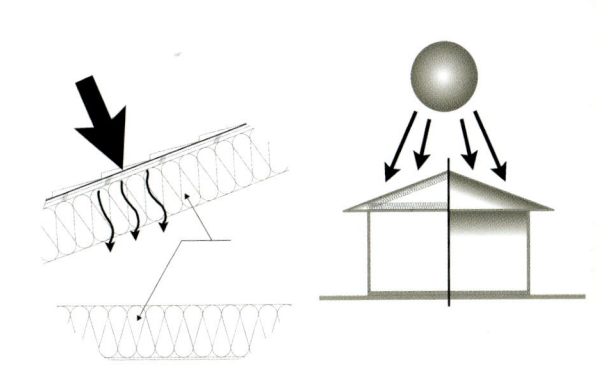

Roof insulation.

energy basically gets trapped inside that airspace and heats up the air there. Every single ridgeline has a ridge vent built into it and then underneath the eaves where you can't see them are inlet vents. What happens is the sun shines on the roof, the air inside that space heats up, becoming hotter, lighter, and lower in density than the surrounding air. Lower-density air rises to the ridge vents and is naturally vented off the ridgeline and cool air is brought in from the eaves. And the whole roof is like that."

Radiant barriers are structural panels installed in attics and walls of a residence to reduce radiant heat gain. The panels can be made from a number of substrates including plywood sheathing, plastic films, air infiltration barrier material, and environmentally friendly wood-based compounds. One or both sides are covered in a highly reflective material, usually aluminum. Radiant barriers are designed to work in ventilated spaces.

Attic Ventilation

For best results in keeping a house cool, cooling down the attic is a priority. Radiant barriers and attic insulation block up to 90% of radiant heat energy from entering the home, but that heat energy has to go somewhere. As we all know, hot air rises (think hot-air balloons) so all that hot air trapped in the attic needs to escape to effectively cool the home. Through proper attic ventilation, the hot air should escape back outside, into the cool trade breezes and not into your home.

There are several types of attic vents that allow that hot air to escape and cool air to enter, and they work together to circulate the air through the attic space. Eave vents and soffit vents are located under the eaves of the roof, where the roof overhangs from the wall of the house. In Hawai'i, soffit vents are called pigeon holes, and you'll often find birds nesting in the round nooks. These vents allow cool, fresh air to enter the attic space. Ridge vents are located at the crest of the roof, under the ridge or highline. Ridge vents allow the hot air to

---

Techshield is sold in standard 4' X 8' panels and will cost you upwards of $30 a panel. Ask the distributor if they offer a 20-year transferable warranty on the product.

A foil radiant barrier—just the reflective metallic sheet—is available in a 4' X 125' roll for around $60. Salvage and reuse scrap plywood or pressboard, attach the foil to the board, and make your own radiant barrier.

escape at the highest point in the roof line. As wind travels over the top of the ridge vent, it literally sucks the hot air out of the attic.

Another option for ventilating the attic is with gable vents. Gables are large rectangular vents in the walls of the attic under the highest point of the ridgeline of the roof. This type of ventilation is most effective if the vents are located on the windward and leeward sides of the house so the trades can easily blow through.

Solar-powered vent fans can help circulate the air and lower the attic temperature dramatically when the house is in a location that does not receive the benefit of the trade winds. Pigeon holes are a must for solar attic fans to achieve the cooling results, so check with a licensed contractor to make sure your attic is properly ventilated.

Thermal Chimney

Just like we learned in basic chemistry class, things move from areas of high pressure to areas of low pressure to reach a level of equilibrium. This is how water and nutrients travel through our cell membranes and how air flows in our atmosphere. These same principles hold true for the way air flows through our homes. When designed properly, our homes can become "thermal chimneys," where cool air naturally replaces hot air to keep the home comfortable and cool without the use of energy-sucking appliances.

Cool air is denser than hot air, so it is heavier and sinks to the ground. Hot air is less dense and rises into the sky. Based on these physical properties of air, it is possible to push hot air up and out of your home with cool air to keep your house comfortable without air conditioning, just like a chimney sucks the hot smoke up and out of the home.

For this natural approach to cooling your home to work, cool air must be brought into the home as low as possible. If your home utilizes post-and-pier construction, vents can be put

---

The trade winds blow about 90% of the time in summer and about 50% of the time in winter. They blow predominantly from a northeasterly direction, but local geography and landscape can alter the direction of the air flow.

—Courtesy of DBEDT

Total vent area should be at least ½ square inch for each 1 square foot of attic area, divided equally between ridge and eave vents.

—Courtesy of DBEDT

Solar-powered gable vents are an option for using the sun's energy to help actively circulate the air in the attic. They cost about the same as a solar attic fan and with state and federal tax credits, they should be in the $350 to $400 dollar range.

HOT AIR OUT HIGH

COOL AIR IN LOW

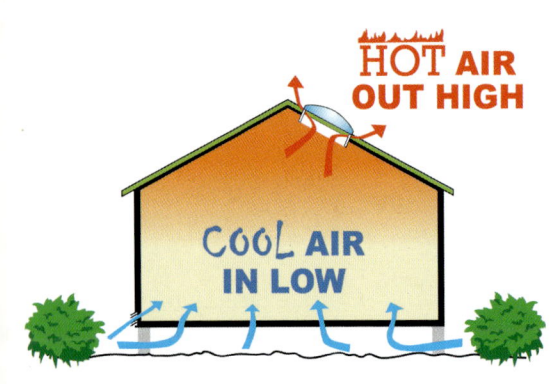

Chimney effect.

Natural ventilation helps reduce health hazards associated with mold and mildew.

To determine how much of a roof overhang to build:

Use a 45° angle for overhangs over windows that face south (sun is lower in the sky in the south).

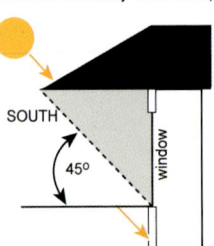

Use a 70° angle for overhangs over windows that face north (sun is higher in the sky in the north).

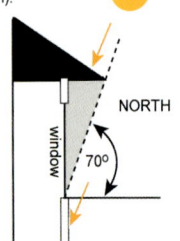

SOUTH 45° window

NORTH window 70°

Roof overhangs.

into the floor to bring cool air in from under the house. Another way to bring cool air into the home is through low windows around the home that have dense vegetation planted close by. Plants naturally cool the air as a byproduct of photosynthesis and that cool, dense air will filter into the house.

Next, it's necessary to have an escape route for your hot air, and that escape route should be as high as possible. Possible escape routes are very high open windows, ventilated skylights, ridge vents, gable vents, and exhaust fans. With the movement of cool air past any of these openings, the hot air will literally be sucked out into the cooler air passing over your roof, leaving your attic space—and your house—cooler.

Shading

Just as the roof can get hot and transfer that radiant heat energy into the house, walls can get hot, heating up the house from the sides in, instead of the top down. The best thing to do to keep the walls cool is provide shade. Shading is accomplished a few different ways. Overhanging eaves are a great way to shade walls and windows that take a direct hit from the sun (most people can attest to the benefit of a large overhanging eave over west-facing windows to block the harsh and hot afternoon sun). Lanais are a great way to shade walls and create a space for relaxing, conversation, and family. Landscaping provides shade to house walls and cools the surrounding air naturally.

Even if shade isn't an option, radiant barriers or insulation sure are, and both do the job of keeping that radiant energy out of your home. Don't forget: white paint reflects the sun and improves cooling performance of these design basics.

Natural Lighting
Windows

In Hawai'i we are fortunate to have warm and abundant sunlight all year long. Of course we have our rain squalls and clouds, but they usually pass overhead quickly with the push of the trade winds. The windows in our homes are a very important yet often overlooked feature of a green home. They are a gateway for cool trades to pass through the abode and allow the natural daylight to shine into our lives, reducing the need for electric lighting. But there is one downside to the passive nature of windows. Solar heat can easily radiate through the glass and into our homes, creating a need for additional cooling—a strain on appliances and the checkbook. The sunlight also brings in harmful UV rays that damage and fade household items like carpet and paintings.

Let's start with the natural ventilating properites of the window. There are several different types of opening windows that allow the breeze to enter our homes, and the way the windows open directly correlates to the amount of air that is able to enter and circulate through the house. Compared to other types of windows, casement windows that open like a door on a hinge can be opened almost completely and allow the most air to pass through. Awnings, where the hinge is located at the top of the window and the window opens outward, and jalousies are the next best bet for the amount of air that can pass through the window area. Sliding, single-hung, and hopper windows only allow about 45% to 50% of the total window area to be open.

There are many ways to shade a window that block the sun's heat from radiating into the home—overhangs, awnings, trees—but sometimes none of the methods are applicable. In this case, special windows can be purchased that are coated in reflective film to block the radiant heat from entering the home,

> **How much roof overhang is necessary to do the job?**
>
> Use a 45-degree angle for overhangs over windows that face south because the sun is lower in sky in the south.
>
> Use a 70-degree angle for overhangs over windows that face north because the sun is higher in the sky in the north.

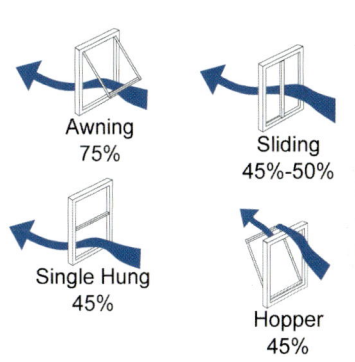

Casement 90%
Awning 75%
Sliding 45%-50%
Jalousie 75%
Single Hung 45%
Hopper 45%

Window options.

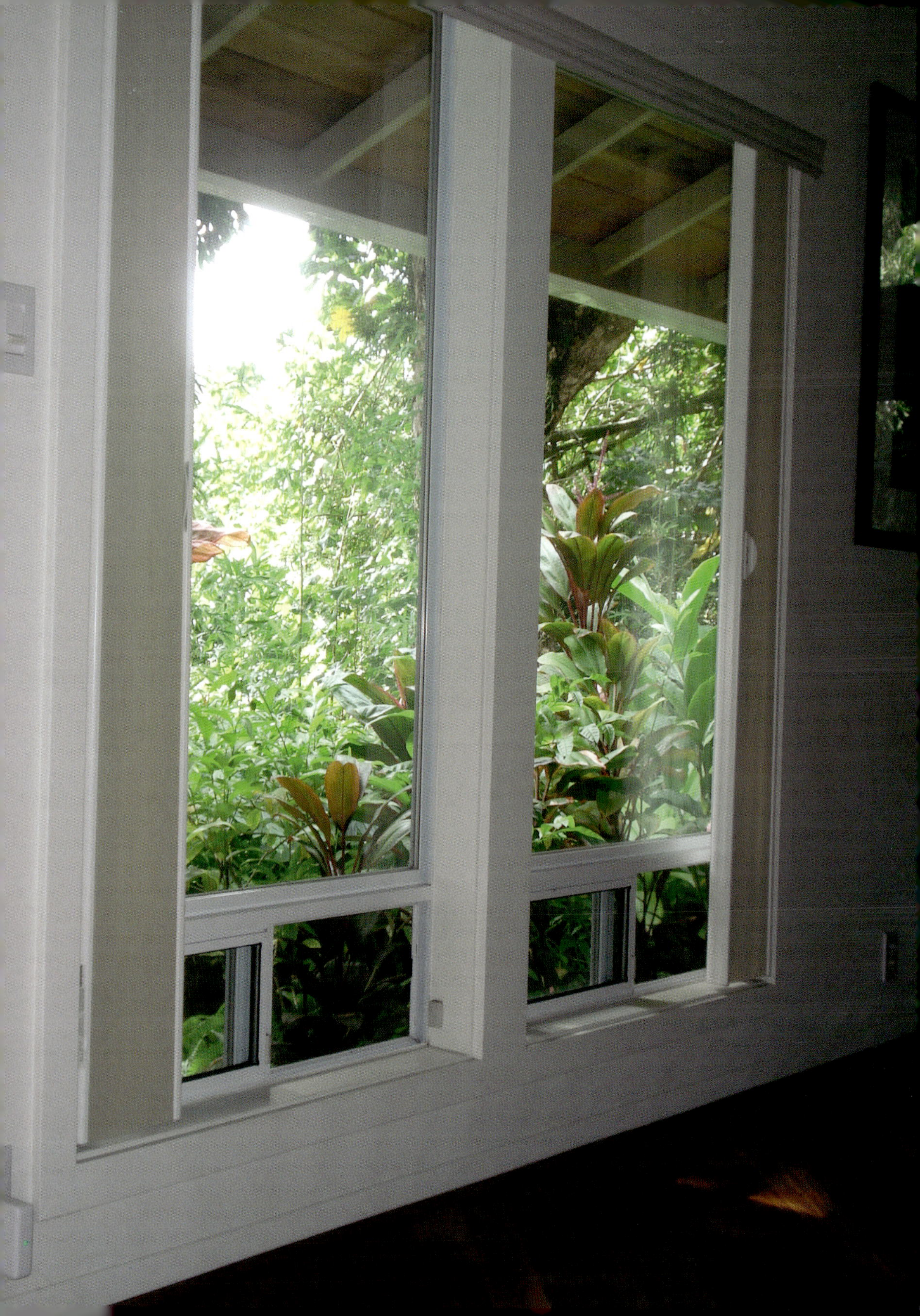

yet allow light to pass through. These windows are called Low-E (low emmitance) windows.

The National Fenestration Rating Council (NFRC) partnered up with the Energy Star program and developed a window rating system to determine efficient window properties. All Energy Star-qualified low-E windows carry a label with the window's rating, which takes into account several factors: U-factor, Solar Heat Gain Coefficient (SHGC), and Visible Transmittance (VT).

The U-factor indicates the rate of heat loss. The lower the U-factor, the greater the window's resistance to heat flow. In Hawai'i, a U-factor lower than 0.65 is recommended.

The SHGC is the most important factor for windows in warmer climates like ours. It is the fraction of solar radiation that passes through the window. The lower the SHGC, the less solar heat it transmits. It is recommended in Hawai'i to use Low-E windows with a SHGC lower than 0.40.

The VT is a measure of the visible light transmitted through the window. The higher the VT, the more light that passes through the window. Select windows with a higher VT to maximize daylight and view.

Low-E windows are double-paned windows and the coating is actually applied to the inside of the panels of glass. If you already have double-paned windows, it is possible to keep the existing frames and just switch out the untreated panes of glass for the low-E panes. If you have single-pane or other types of windows, then it is necessary to replace the window frame as well as the windows.

Milgard Windows is a window manufacturer that produces Low-E windows that carry the Energy Star seal of approval and are readily available in Hawai'i. Milgard Low-E windows come set in vinyl frames, either tan or white, and are reasonably priced. A six-foot wide by three-foot high, horizontally sliding Low-E window with vinyl frame costs a little above $300. Prices vary depending on the dealer, of course. Most deal-

---

That's a lot of abbreviations and figures for anyone to handle. The U.S. Department of Energy's Windows and Glazings Program and the Efficient Windows Collaborative have put together worksheets, charts, definitions, and explanations specific to Hawai'i about Low-E and other types of efficient windows on their website, www.efficientwindows.org.

Skylights can provide a wealth of natural light to illuminate and enrich a home by brightening up those dark spaces like kitchens, bathrooms, and hallways. There are skylights availbable that open for ventilation, so your skylight can have a hand in both lighting and ventilation.

> Vented skylights should only be used if the home does NOT use air conditioning.

ers also charge an additional crate fee. Milgard windows come five to a crate. Whether you buy one window or five, you'll be charged the same fee.

Solar Light Tubes

It seems like there's at least one dark corner in every house. Maybe it's the hallway, or a bathroom without a window to call its own, or how about the room plagued by a single window on the north side of the house. It's so dark in there, not even orchids will grow. Now there's a quick and relatively inexpensive way to get a lot of light out of a little package. Solar tubes, also called tubular skylights, are built specifically to accomplish this goal.

Solar light tubes are just that—tubes that extend from the roof to the ceiling and bring natural light into the home. They are usually ten to thirteen inches in diameter. The top of the solar tube is capped with a clear plexiglass dome that captures light coming from any direction, at any time of day, and focuses it into the tube. The tube itself is made of highly reflective material and the light reflects and refracts thousands of times as it travels through the tube. Before the light exits the tube and enters your home, it refracts one more time through hundreds of

Solar tubes.

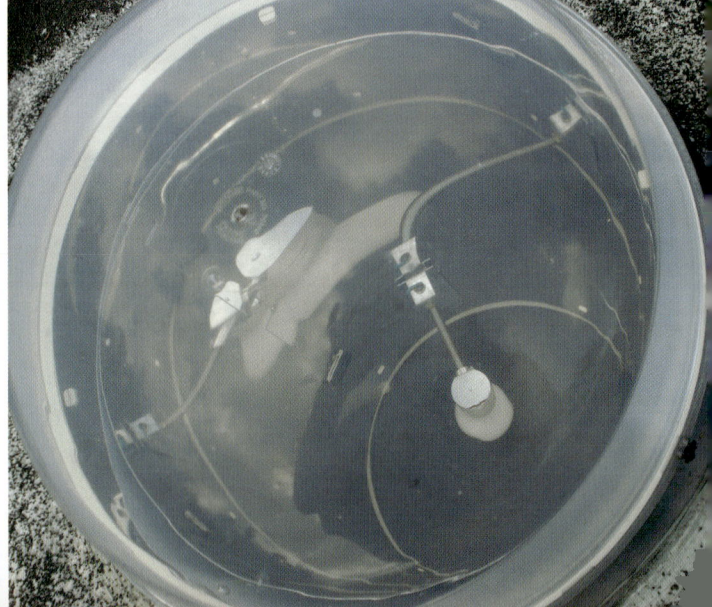

prisms in the plexiglass cap that you see from inside your home, literally throwing light into the living area.

"A ten inch light solar tube gives off 600 watts of light power. It's amazing. It actually collects the light, draws it in, magnifies it and then shoots it out," says Ray Heitzman, owner of Ray's Solar Fans on Oʻahu. "It can be a completely overcast, rainy day, and it will still give off lots of light—not as bright as it would be with the sun out, but it will still be bright enough to light up a hallway."

The beautiful thing about solar light tubes is that they brighten up your home with free, natural light without dispensing heat into the home. The plexiglass dome on the roof is designed to capture the light and at the same time, reflect the heat. During the day, a solar tube literally takes the place of an incandescent or fluorescent light.

"A regular light will have a line of demarcation where it's dark and you can see a shadow. There is no shadow with light from a solar tube; all it does is give pure light in all directions," explains Heitzman. "You get this warm, very powerful, but not overwhelming light just filling the entire area that needs light. I've put them into bathrooms and the light just spills out into the hallway. It's one of the most amazing, untouched inventions around."

Laurie Carlson decided to put in a solar light tube while her solar contractor was up in the attic installing her solar attic fans. "We had a solar tube installed right above my husband's desk and during the day it reflects light down through the tube and lights that whole area of the house. At night, there's a compact fluorescent light we can turn on in there. It was a real dark area and now, during the day, it's like having an extra window."

Solar light tubes start at about $600 for a basic setup, but when you never need to flip a light switch during the day, the energy savings over the long haul are well worth the initial investment. And don't forget, the state of Hawaiʻi offers a 10% tax credit on solar light tubes installed by a certified solar contractor.

Solar light tubes are available with a fluorescent light inside the tube for night lighting, totally replacing the need for an incandescent bulb. They are also available with an exhaust fan setup for bathrooms.

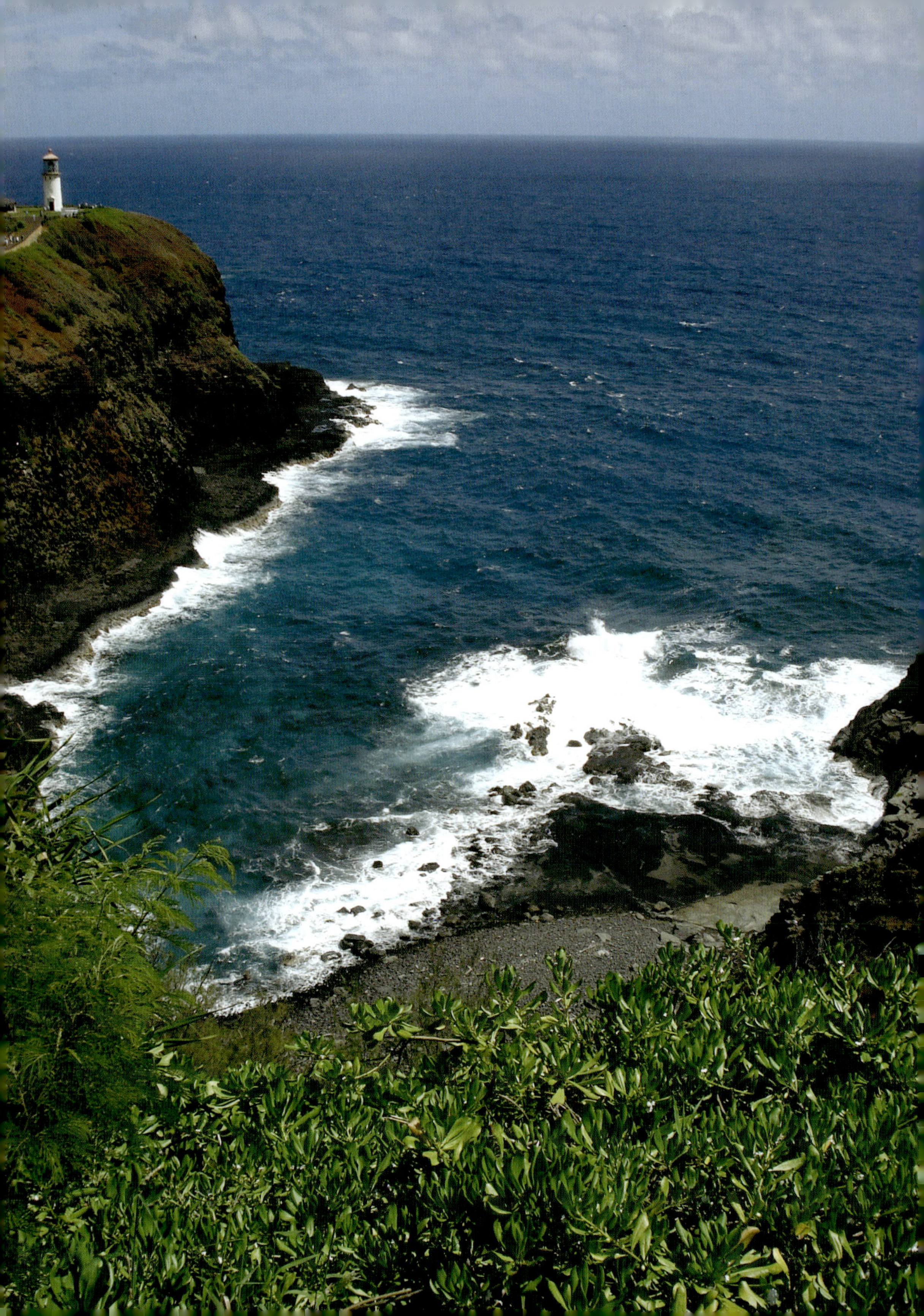

three: indoor water conservation

Hawai'i's fresh water is a precious resource, a sentiment as recognized and revered by the ancient Hawaiians as it is today. Without fresh water on our islands, life would not be possible. In pre-contact Hawai'i, fresh water was a gift form the *akua* (gods); nobody owned it. The ancient Hawaiians knew fresh water was the key to life and prosperity, so they lived near perennial streams and springs. They designed extensive agricultural systems with canals, ditches, and terraces to flood the lo'i and create fishponds. Conservation was inherent in their use of the land, and they took no more than they were able to immediately use because they understood the relationships and connections between the people and their natural environment.

Lili'uokalani Botanical Gardens.

If you have fish as pets, then no doubt, you have to clean the tank from time to time. Instead of throwing that water down the sink, use it to water house plants. The nitrogen and phosphorus in the water from the fish waste will be a nutritious meal for your greenery. If you don't have house plants, use the water outside in the garden.

Ko'olau mountains.

When you understand where our water comes from, you'll realize the importance of its conservation as well. As ocean water evaporates under the heat of the sun, northeast trade winds push the moisture toward our island chain. When the moisture-rich air comes up against a mountain range, the air is forced upward. As the air rises, it cools; clouds form and the moisture in the air condenses. Soon it becomes too heavy to remain in the sky and falls to earth as precipitation or rain. At this point, the journey of water has only begun. Once the rain reaches the ground, some of it nurtures vegetation, some is carried away in streams and rivers to the ocean as runoff, and some seeps into the ground, where it percolates though the soil and the volcanic rock, into natural fresh water aquifers below the island.

The journey of a raindrop from the top of the Ko'olau Mountains on windward side of O'ahu down to the fresh water aquifer can take as long as 25 years. Along the way, the water is purified as it percolates, yielding just about the cleanest water on the planet. The water is then drawn up through an underground network of water tunnels, shafts, and wells by the water utility and pumped to reservoirs that send water uniformly to our neighborhoods and homes, ready to drink.

It's easy to take resources like fresh water for granted when we don't need to think twice about a glass of water, flushing the toilet, or watering the garden, but that aquifer is a finite resource and can become tainted by the salt water that it floats atop if the fresh water supply is sufficiently depleted. That is why water conservation is so important to sustaining our quality of life and ensuring our future prosperity.

Water conservation is founded on common sense and simplicity: use only what you need. You probably already use many water-saving techniques on a daily basis without knowing, but there is always room for improvement, and education is the key. Read on for helpful hints, ideas, and money-saving ways to save water and in turn, benefit Hawai'i.

Kitchen

Choose an Energy Star-qualified dishwasher that saves water with every cycle. Scrape the plates to discard food instead of rinsing before putting dishes in the dishwasher. Newer dishwashers are designed to handle the mess, and if you're going to rinse your dishes first, you might as well just wash them. Wait for a full load of dishes before running a cycle.

If you wash dishes by hand, fill up one side of the sink with soapy water so that water isn't running while you wash. If you don't have side-by-side sinks, try filling up a bowl or cup with soapy water. After you've washed all the dishes, rinse 'em. Consider installing a faucet aerator or spray tap that mixes air with water to cut down usage.

Defrost food in the refrigerator instead of running it under water.

Check regularly for leaks under the sink. Check the refrigerator and any ice-making or filtered-water device for leaks as well.

Across the Hawaiian Islands, the amount of precipitation varies widely, from 20 to 300 inches a year, depending on location. In an average year, 2 billion gallons of rain falls on Oʻahu in a single day.

—Courtesy of the HBWS

Bathroom

When taking showers, take as short a shower as possible. Every minute you trim off shower time saves from three to six gallons of water. Try turning off the water while you soap up or shave. Water-efficient shower heads cut down on excess water usage.

Always turn off the water while brushing your teeth or shaving. If you leave it running, you could waste up to one gallon a minute.

Check the sink for any leaks or pesky faucet drips and fix them immediately.

Toilets are one of the main water users in the home. Put a plastic container or toilet dam in your tank to offset the amount of water used per flush. And don't use your toilet as a trashcan. Throw away tissue instead of flushing it. Old toilets use up to five gallons per flush, so your best bet is to invest in an ultra-low flush toilet. The Honolulu Board of Water Supply is offering rebates to cut the cost. Check their website for details.

Laundry

Old clothes washers use up to 40 gallons of water per wash. Switch to a new Energy Star-qualified clothes washer for maximum water efficiency. Use the proper water level setting for the amount of clothing to be washed. If there are no settings, wash with a full load of clothes. Note that front-load washers use less water than top-load washers.

Use the shortest cycle necessary to clean your clothes, and pre-treat stains.

Check hoses often for leaks and repair when necessary.

Leaks

To check if your toilet is leaking, first put ten drops of food coloring in the toilet tank. Don't flush; just wait fifteen minutes. If you see food coloring in the bowl, you've got a leak.

If you have a leaky faucet, you might be wasting more water than you think. A slow drip can waste up to 15 gallons per day and a 1/8-inch stream wastes up to 400 gallons a day. If the handle is turned completely off and the water is still dripping, you probably just need new washers—a quick and cheap fix.

To check your plumbing for major leaks, first turn off all water faucets, pipes, and water-using fixtures so that no water is running; then locate the water meter for your residence. Open the cover and check out the dial. If it's moving, you've got a leak. A leak of only 1/5 gallon a minute wastes more than 8,500 gallons a month. That's a lot of wasted water and a lot of wasted money.

Water Utility Contacts

Oʻahu: City and County of Honolulu Board of Water Supply, Water Waste Hotline (808) 748-5041, www.hbws.org

Maui, Molokaʻi, Lānaʻi: County of Maui Water Supply, Operations Trouble Calls (808) 270-7633, www.mauiwater.org

Kauaʻi: Department of Water, Trouble Calls (808) 245-5444, www.kauaiwater.org

Island of Hawaiʻi: County of Hawaiʻi Department of Water Supply, Customer Service (808) 961-8060, www.hawaii-county.com/water/DWS-main.htm

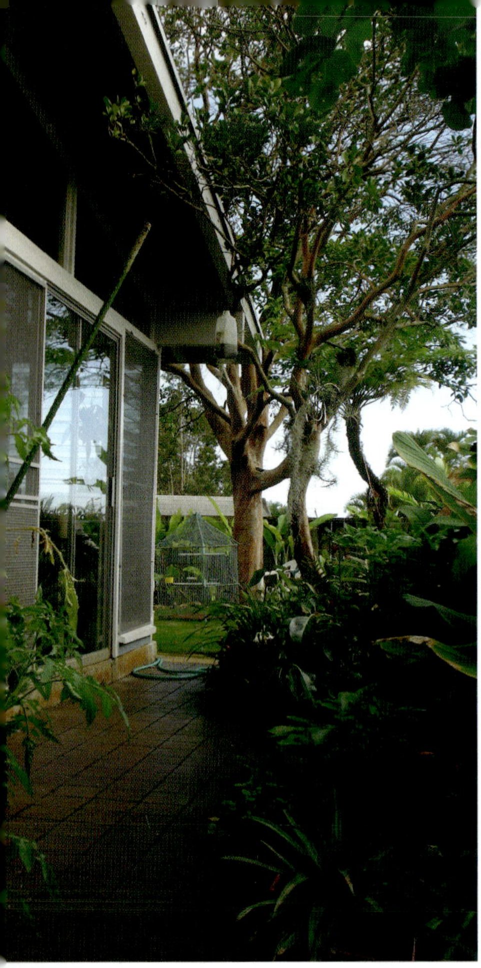
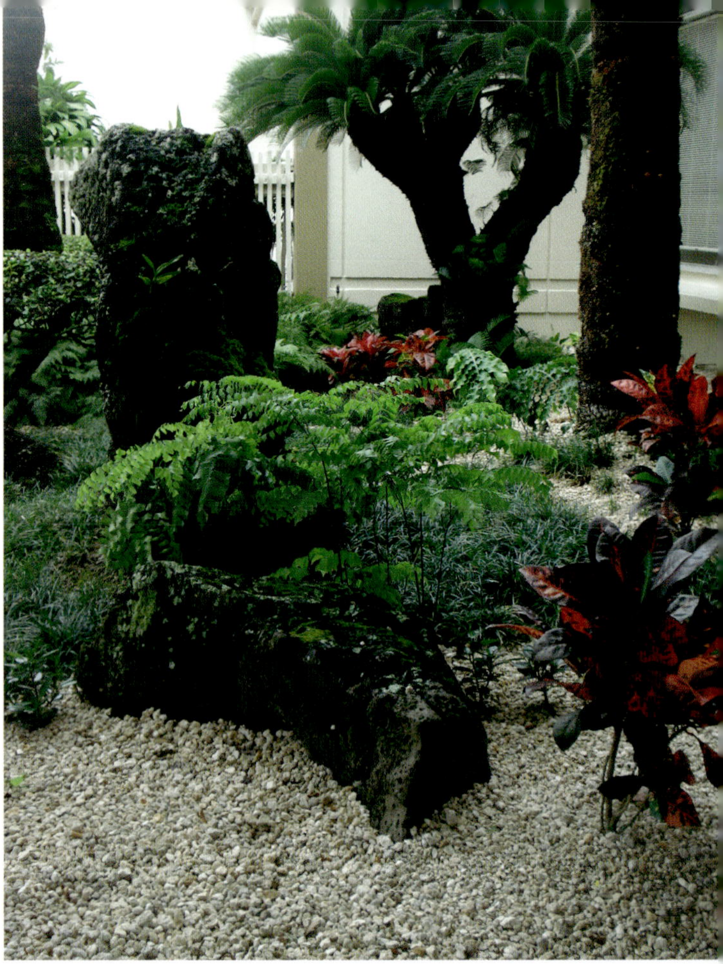
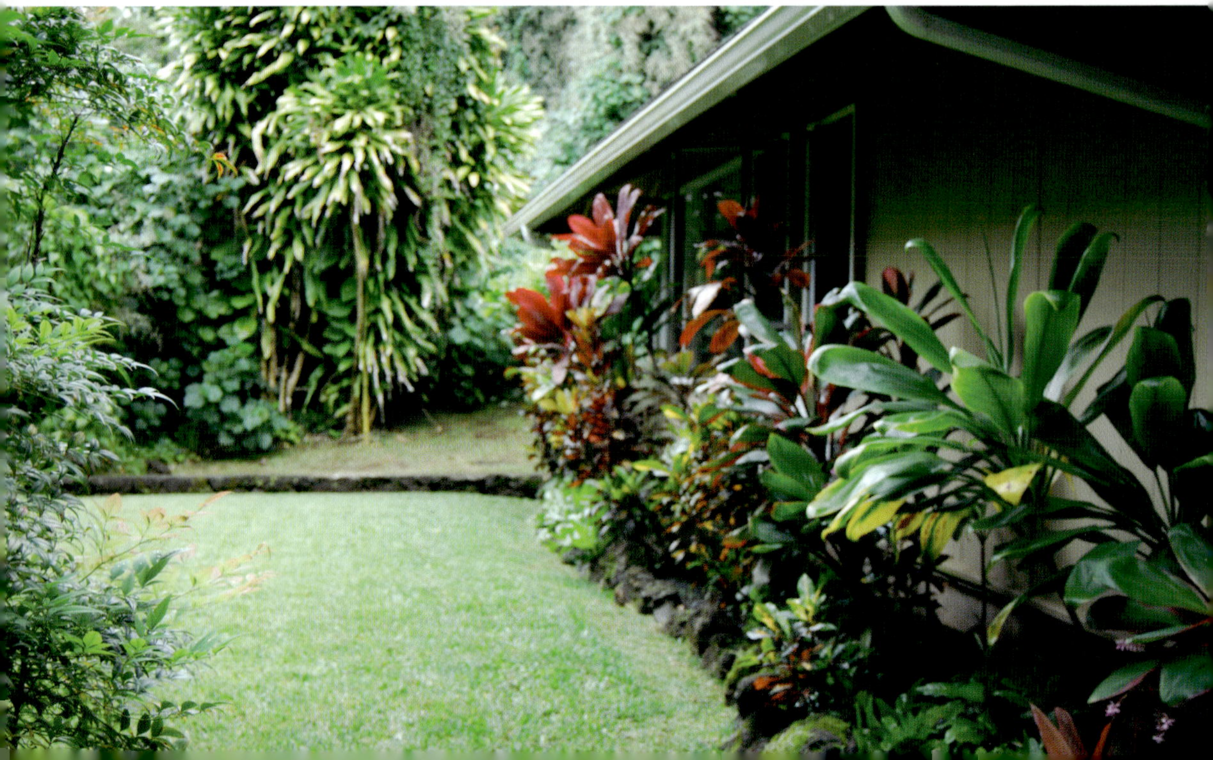

IN THE YARD

part 2

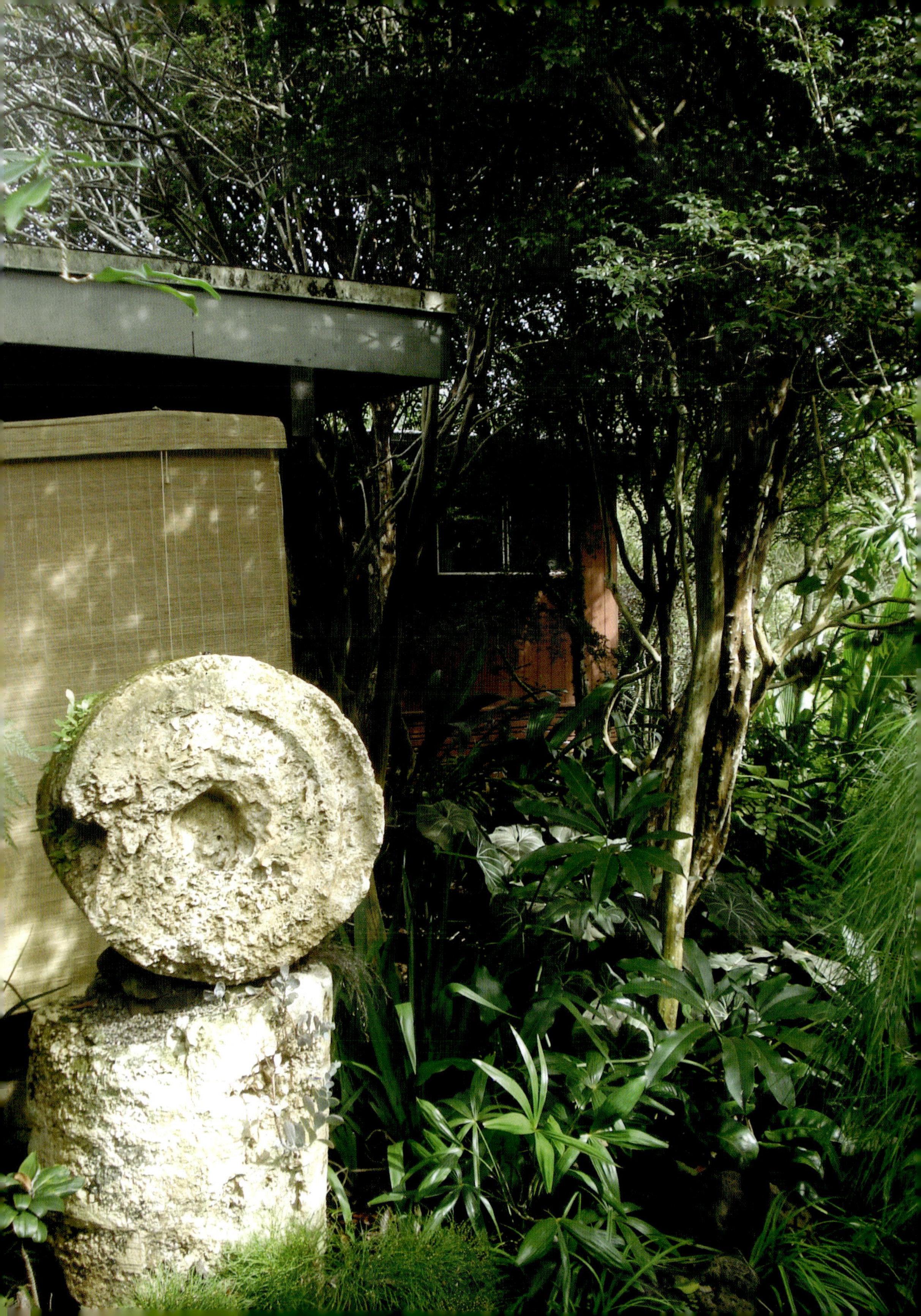

four: landscaping

The area surrounding your home is just as important as all those wonderful things that fill the living space inside the walls and the hard work, diligence, and thoughtfulness necessary to make it comfortable and energy efficient. And when we open the lanai door and step outside, the need for sustainability and energy efficiency isn't fleeting like a bird's song. On the contrary, a green home continues to take shape outside and around the home as well.

Plumeria.

The plants you choose to fill your yard have a significant impact on the shading and cooling of your home and the health of the Hawaiian Islands. Planting native Hawaiian plants that are specific to the climate that you live in will not only beautify your yard, but also cut down on the need for watering and give sanctuary to Hawaiian birds, butterflies, and snails that long for the natural landscape.

There are ways to care for your yard that will not pose a toxic threat to the environment, and great ideas for recycling food and garden waste, creating some of the best fertilizer on this planet.

Water conservation is as big a priority in the yard as it is in the home, and there are many ways to cut your water usage. You can take advantage of the rain and capture rainwater or use recycled gray water to care for your lawn and garden.

But above and beyond all these benefits is the fact that you'll be creating a calm and positive outdoor environment to use for your personal benefit, morning, noon, and night. In addition, the plants and trees play an integral role in the shading and cooling of your home, which contributes to a more comfortable home and curtails energy consumption. Not to mention, the therapeutic effects of a healthy garden can be contagious. Relax and enjoy.

Outdoor Water Conservation

Depending on where you live in the Hawaiian Islands, the climate and precipitation can change drastically. If you live near

Verdant front yard and lanai.

Mt. Waiʻaleʻale on the North Shore of Kauaʻi, then you're privy to the lush greenery that takes advantage of the heaviest precipitation in the world. If you live in ʻEwa, then you reside in one of the hottest and driest spots on the island of Oʻahu. Either way, it is possible to conserve water outdoors and still have a beautiful area to drink in our beautiful Hawaiian weather.

If you do happen to reside in an area with moderate to heavy rainfall, it's still necessary to conserve water. For instance, the average outdoor faucet, the one your hose is attached to, can release up to 500 gallons of water an hour. That is a very generous volume of water. The problem here is that if you forget about the hose running outside, watering the lawn while you're having a Sunday snoozer, the potential for waste is immense. Rain or no rain, it's critical to conserve water.

With some common sense and basic gardening knowledge, you'll have a terrific-looking yard, and you can use less water to get it that way. And we all know that using less water means extra savings on the water bill. Here are some simple ways to conserve water in the yard:

Pressure nozzle.

> Instead of putting your finger over the end of the hose to create a wasteful, makeshift sprayer, use a cutoff or pressure nozzle on the end of the hose. These nozzles allow you to use only the amount of water you need at that time, and enable you to direct it exactly where you want the water to go. Pressure nozzles are relatively inexpensive and some come with a lifetime guarantee. There are adjustable nozzles that let you chose the type of spray, from soakers to showers.

> Lawns do not need to be watered every day. In fact, you shouldn't water your lawn until it shows signs of

Patricia S. H. Macomber wrote the bible on rainwater catchment systems for Hawai'i, aptly named Guidelines on Rainwater Catchment Systems for Hawai'i. Based on her work at the College of Tropical Agriculture and Human Resources at the University of Hawai'i at Mānoa, the publication is a must-read if you're one of the 50,000 people on the Big Island who rely on rainwater for all their water needs, or if you'd like to get started on your own system. A free download of the publication is available at http://www.ctahr.hawaii.edu/oc/freepubs/pdf/RM-12.pdf.

needing a liquid boost. If you step on the lawn and it bounces back, no need to water. If it stays flat, time to water. A good rule of thumb is to water less often, but with a good soaking. It promotes deep root growth and healthier grass. If you stick your finger into the soil and it's dry, then a good soaking is in order. Runoff should never occur when properly watering the lawn.

When it's time to cut the grass, set your mower to a higher setting. Longer grass does better in the heat and shades the soil, keeping it cool and preventing evaporation.

Always water in the mornings, before 9 AM, and in the evenings after 5 PM to prevent evaporation from direct sunlight and the heat of the midday sun.

Take notice of the weather conditions before you water. If it's windy, don't use sprinklers or misters; the majority of the water will be swept away by the wind. Instead use soakers or spray settings with larger drops of water and keep the nozzle close to the base of the plant or grass. If it's about to rain, no need to water either; take the day off of garden duty, the rain will take care of the yard for you. Remember to turn off any automated irrigation systems when it starts to rain. Some automated systems come with rain-shutoffs or soil moisture sensors to help with water conservation.

Put mulch or grass clippings around the base of trees and shrubs to prevent water evaporation from the soil and prevent weeds.

If you have a soggy spot in your yard or a patch of grass that grows ten times faster than the rest of the lawn, then most likely you have an underground leak. This is a

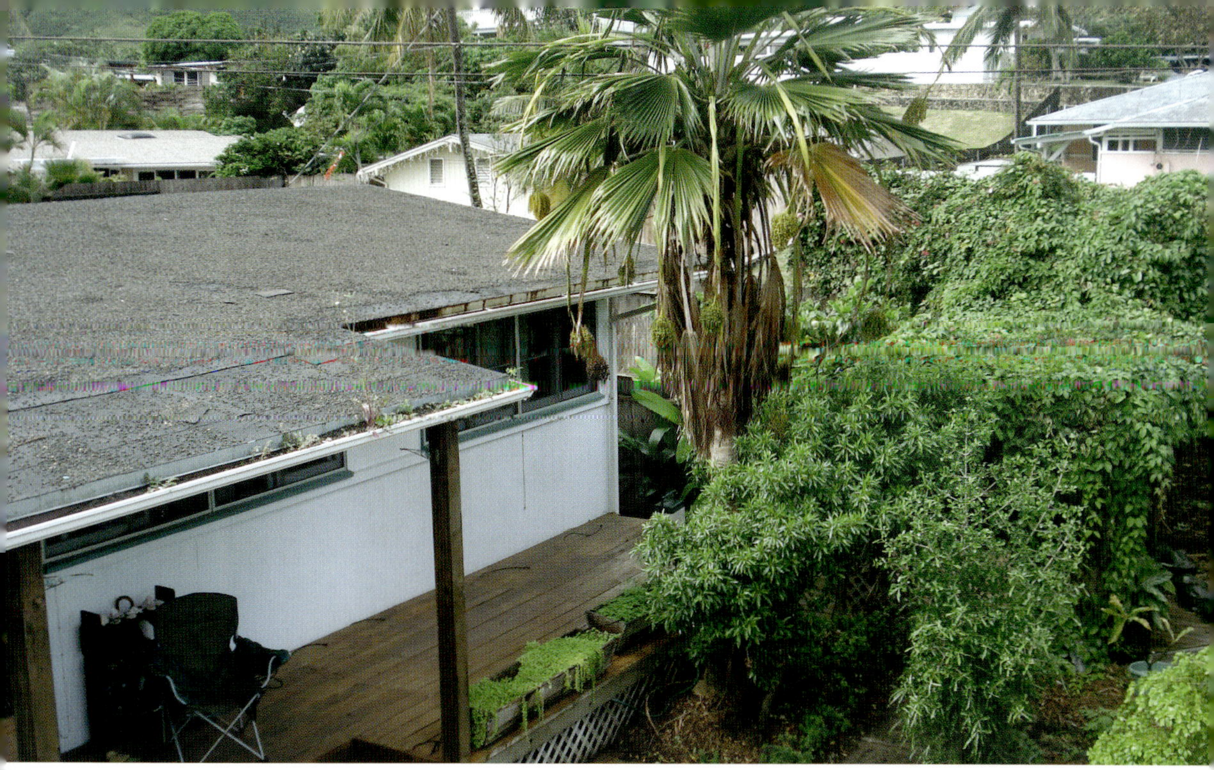
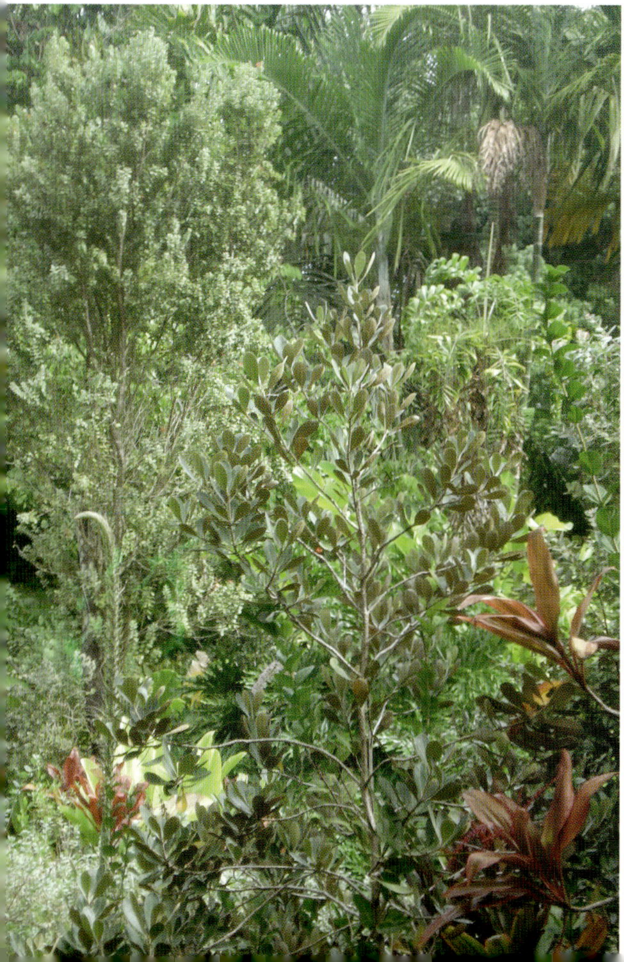
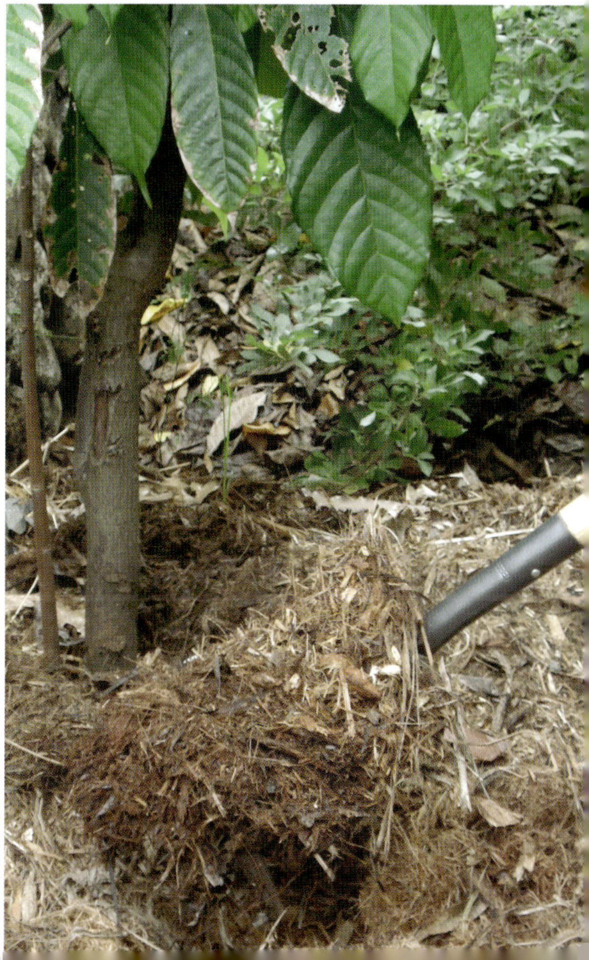

Leaky outdoor faucet.

tremendous waste of water and these leaks just get bigger with time. The only solution is to locate and repair leaks immediately.

Drip irrigation systems save water by putting water directly at the base of the plant where it can be absorbed directly by the roots. The slow drip soaks the soil thoroughly and keeps evaporation to a minimum.

Apply only the minimum amount of fertilizer necessary to the do the job, as excess fertilizer can leach into runoff water that travels to streams, rivers, and oceans. Fertilizers are extremely harmful in these habitats, accelerating algal blooms and even killing coral reefs.

Use safe pesticides instead of toxic ones to take care of garden pests. For common pests like white flies, scale, and aphids, spray flowers and leaves weekly with a solution of dishsoap and water.

Sweep up sidewalks, walkways, and driveways instead of hosing them down with water.

Attach a rain gutter to a 50-gallon plastic drum and store rainwater in your own rainwater catchment system. Harvesting rainwater for use in your yard and garden will keep you from turning on the hose as often.

Trees for Shade

Trees are amazing creatures. They have inhabited the earth for about 360 million years, long before flowering plants were around (they came about 200 million years later). There are over 50 families of trees and hundreds more individual species. Trees grow in all shapes and sizes, from tall and narrow like the Cook Island Pines (often mistakenly called Norfolk Pines)

Mature shade tree.

that dot the islands, to the shorter and squat monkey pods that spread their feathery canopy across the sky.

Trees are our steadfast neighbors. They clean the air and provide us with the oxygen we rely on for survival. They can provide food and materials and shade us from the sun, cooling the air beneath their limbs and leaves. They give shelter to animals and provide climbing perches for nimble keiki and avid tree lovers. In rainforests the broad leaves of canopy trees shield the ground and a thin layer of delicate soil from the erosive power of falling raindrops, stabilizing the earth.

Trees are incredibly wonderful living resources for shading your home, windows, walls, or lanai from the sun. Trees can also do double duty and act as wind, noise, and visual screens. But selecting a tree to get the job done is not as easy as going down to the local nursery and grabbing just any old tree to plant in your yard. Considerations must be made and patience employed if you are to plant the right tree for the site to get the job done properly.

> Native Hawaiian plant species occurred naturally in Hawai'i before humans arrived. Endemic Hawaiian species are found only in Hawai'i and nowhere else in the world. Alien or exotic species were introduced to the Islands, and invasive species are alien species that grow rapidly and spread easily.

> Polynesians brought 24 plants with them in their canoes including *kalo* (taro), *ki* (ti) and *niu* (coconut palm).

Kevin Eckert, ISA (International Society of Arborists) certified arborist and consulting arborist for Arbor Global, suggests a simple, outlined approach to selecting the right tree, so grab a pen and paper and follow along.

1) Determine the function of the tree.

The function of the tree is the job you are trying to accomplish, like shading, a wind screen, or a noise screen. In terms of shade, make sure you know what you are shading: a roof, a wall, an air conditioning unit, or a window. Determining the function of the tree will help you determine the characteristics of the tree you are looking for. For example, let's use shade as our desired function.

2) Determine the structure of the tree.

Now that you have decided what it is you want to accomplish by planting a tree, we need to look at the structure, which is the architecture or form of the tree. If you need the tree to shade your roof, choose a large and spreading species. If you want a wall of your house shaded, you want something shorter and more columnar. Once you know the form of the tree you need for the job, then take into consideration fruit and other aesthetic properties.

3) Make a list.

Now that you have determined the function and structure of the tree, make a list of all the trees you would like to plant in your yard to accomplish your goal. Include as many native Hawaiian species as possible. If you need some help getting your list started, Kevin Eckert has a suggestion: "It's a nice practice to wander around in the neighborhood where you live and see what's working, see what's growing there. Trees are very site-specific [to the area's] soil, light, water. One way to get a good understanding of what will work is to take a walk around and see what's planted, see what looks good and looks kind of like what you're looking for."

4) Determine the site space.
This is a huge and often overlooked consideration that, if not adhered to, can lead to costly maintenance and a big headache years down the road. Look up! Make sure there are no overhangs or utility lines overhead. This is referred to as the lateral airspace. But here's the tricky part: you need to gauge the lateral airspace needs for the tree at maturity, not for what it looks like when you bring it home from the nursery. Now keep in mind, your tree might not have a mature crown for 15 to 20 years. "If you're planning on planting a big spreading tree like a banyan or monkey pod, don't put it in a 20-foot wide area," warns Eckert.

ʻŌhiʻa lehua.

Next, look at the surface area on the ground and make sure the tree has room to spread. Are you planting too close to a sidewalk, pool, or other obstacle that could be damaged by the tree further down the road?

Finally, think about what's under the ground. "The roots hold the tree up, they provide nutrients and water, they provide storage, and they transport those materials up into the tree," explains Eckert. "If you restrict the root system, you could have major problems, and this could happen 10 or 15 years from now." Underground utilities are also a major concern for spreading roots.

5) Determine the planting conditions.
This is basically a more in-depth look at the site. How much sun does the site receive and at what times during

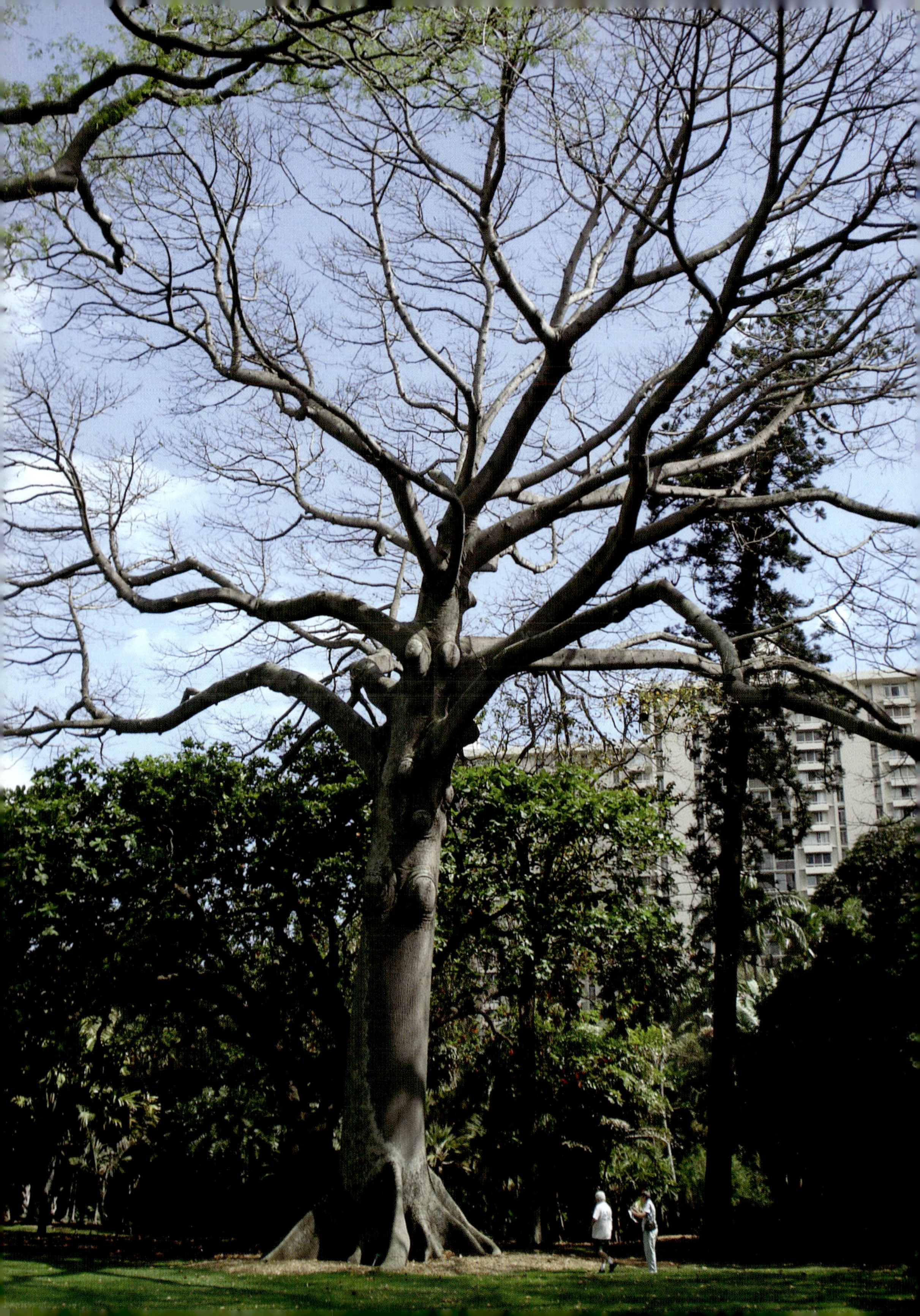

the day? What type of soil will the tree be growing in? Is the location prone to strong winds? Is the home located close to a beach where salt water spray or salt water in the ground water could be a concern?

6) Determine the required maintenance.
Keep in mind that if you plant a tree over a pool, you're probably going to be cleaning the pool often, removing soggy leaf debris. Landscape architect Leland Miyano suggests keeping maintenance to a minimum. Trees with moderate growth are stronger and shed leaves and branches less often. Fast-growing species are more brittle and constantly lose branches and leaves to the elements.

7) Narrow your search and pick your tree.
Based on your site, the planting considerations, and the function and structure of the tree, start crossing trees off the list that just don't seem to work. The goal is to look at the natural planting conditions on the site and then select a tree that is going to adapt and thrive, not just survive. Always avoid species with invasive properties.

Leland Miyano has designed gardens in Hawai'i from single-family homes to resort hotels and offers this advice: "Anytime you have plants near a house, it's great to cool. It's a principle in good design. Most people have air conditioning and don't even think about plants as a cooling thing.

"When planting trees near a house, you want to see something with moderate growth, something with smaller leaves or leaves that don't shed so much, and it depends on which side of your house the plants are on. Also, you have to figure out the eventual height of the tree. You don't want to have a giant tree when you're only planting in five feet of setback. In that instance you need something vertical, otherwise it's going to be a lot of pruning."

A Pocket Guide to Hawai'i's Trees and Shrubs, by H. Douglas Pratt, is an excellent handbook for learning about native and exotic species and helping you make decision about what to plant in your yard.

How to plant a tree—and have it survive!

You've made your list, narrowed down your choices, and selected the perfect tree for the site to get your shading job done just right. Your work is not over yet, so here is a quick run-down on how to plant your new tree and have it survive.

1) Select a healthy tree.
At the nursery, choose a healthy, structurally sound specimen, one with a nice single trunk. This promotes the growth of a central leader, which will allow your tree to grow up strong and at its maximum rate. Take a look at the foliage and make sure there are no insects, disease, or damage, which can weaken and even kill a young tree. Don't go for the biggest tree in the smallest pot; you're not getting a good deal. Make sure there are no roots poking out of the holes in the bottom of the pot and that there are minimal girdling roots, roots that circle the inside the pot.

2) Dig the pit.
Find the root collar, where the trunk flares into the roots. This will tell you how big the root ball is and how deep to dig the pit. Take the height of the root ball and dig the hole 90% of that height. That's right, smaller than the root ball. This leaves room for the root ball to settle into the ground without going too deep. The root collar should always be above ground to promote healthy growth. In addition to the depth, dig the hole two to three times wider than the root ball. This gives you room to work without damaging the roots. Save the dirt.

3) Install.
Set the tree in the pit and orient the tree properly so that the trunk is vertical. Next, turn the tree so that the prominent limbs, the ones you plan on keeping, do not become obstacles in the future. Backfill a little soil around the base of the root ball to

support it, continue to fill in the hole with a third of the original soil, and then wet it. There is no need to tamp or pack the soil. Add another third of soil, wet again, and then finish off with the last of the soil and water. Fill up the hole to the soil level and do not put soil on top of the root ball, which should be poking up just slightly out of the ground.

4) Mulch it.
Put about a four-inch cover of mulch over the root ball, making sure it does not touch the trunk. If you purchased a tree with a proper-size root ball, there should be no need to stake it unless you live in a windy area. If that is the case, drive the stake in the windward side of the soil, not the root ball, and use broad and flexible ties to put around the trunk, like an old bicycle inner tube. Attach the supports as low as possible on the trunk. Check the points of contact monthly to make sure the ties are not damaging the tree. If done properly, one growing season of staking will suffice.

5) Nourish it.
Watering is the single most important thing you can do to ensure the survival of your new tree. New trees need more water than established trees and irrigating regularly for the first three to six months is necessary. A slow soaking so that the water percolates deep into the soil is best, promoting deep root growth. Be sure not to over-water though, as trees can drown. Use the finger test to check the soil for moisture. Stick your finger in the soil; if it comes out dry, it's time to water. If it comes out soppy wet with soil on it, no need to water. The perfect soil would be moist, but not soggy.

There is no need to fertilize a young tree, or any healthy tree or plant for that matter. Fertilizers can actually harm young trees, attracting insects and diseases that enjoy the fertilizers as well.

Planting Native Hawaiian Plant Species

The Hawaiian archipelago is the most isolated landmass in the world, born of a volcanic origin and slowly eroding back into the ocean. Because of our geography and isolation, thousands of species have evolved and flourished across the Islands over the millennia—plant, insect and animal species found nowhere else on this earth and specialized according to the geographic niche they inhabited. From rainforest to coastal desert, there are about 150 different ecosystems in the Hawaiian Islands. Just about the only ecosystem not represented is the frozen tundra.

In pre-contact Hawai'i there were over 1,400 endemic species of plants. Today, Hawai'i is unfortunately the extinction capital of the U.S.: around 10% of the native flora is extinct; about 180 plant species are listed as endangered and rare (50 or fewer individuals remaining in the wild); two-thirds of the native bird species are extinct; more than half of the endemic land snails are extinct; and the majority of all the coastal and lowland plant species have been wiped out for quite some time due to human occupation, development, and invasive species. Beginning with the ancient Hawaiians' use of lowland areas for farming and perpetuating at an expanding rate over the last 200 years, Hawai'i's fragile ecosystems have fallen prey to the devastating effects of blooming human populations and introduced alien species (pigs, rats, cattle, weeds, etc.). Now, sharp vertical cliffs and high peaks, locations difficult for humans and alien pests to access, are some of the only remaining natural places where native species still reside.

Recently, much attention in the scientific and horticultural community in Hawai'i has been paid to the decline and extinction of native plant species, especially in populated lowland areas and watersheds. Efforts have been made to identify and collect seeds from endangered and threatened native plants in the wild for use in the propagation and reforestation of native Hawaiian plants on the verge of extinction.

To demonstrate the threat of invasive species, take for example the gall wasp from Africa that has nearly brought the entire wiliwili species to extinction. With no concrete solution for eradicating the tiny wasp, around 90 pounds, or 66,000 seeds, have been collected and preserved in a seed bank at the Lyon Arboretum on O'ahu. According to junior researcher Alvin Yoshinaga, the seeds should remain viable for up to 30 years.

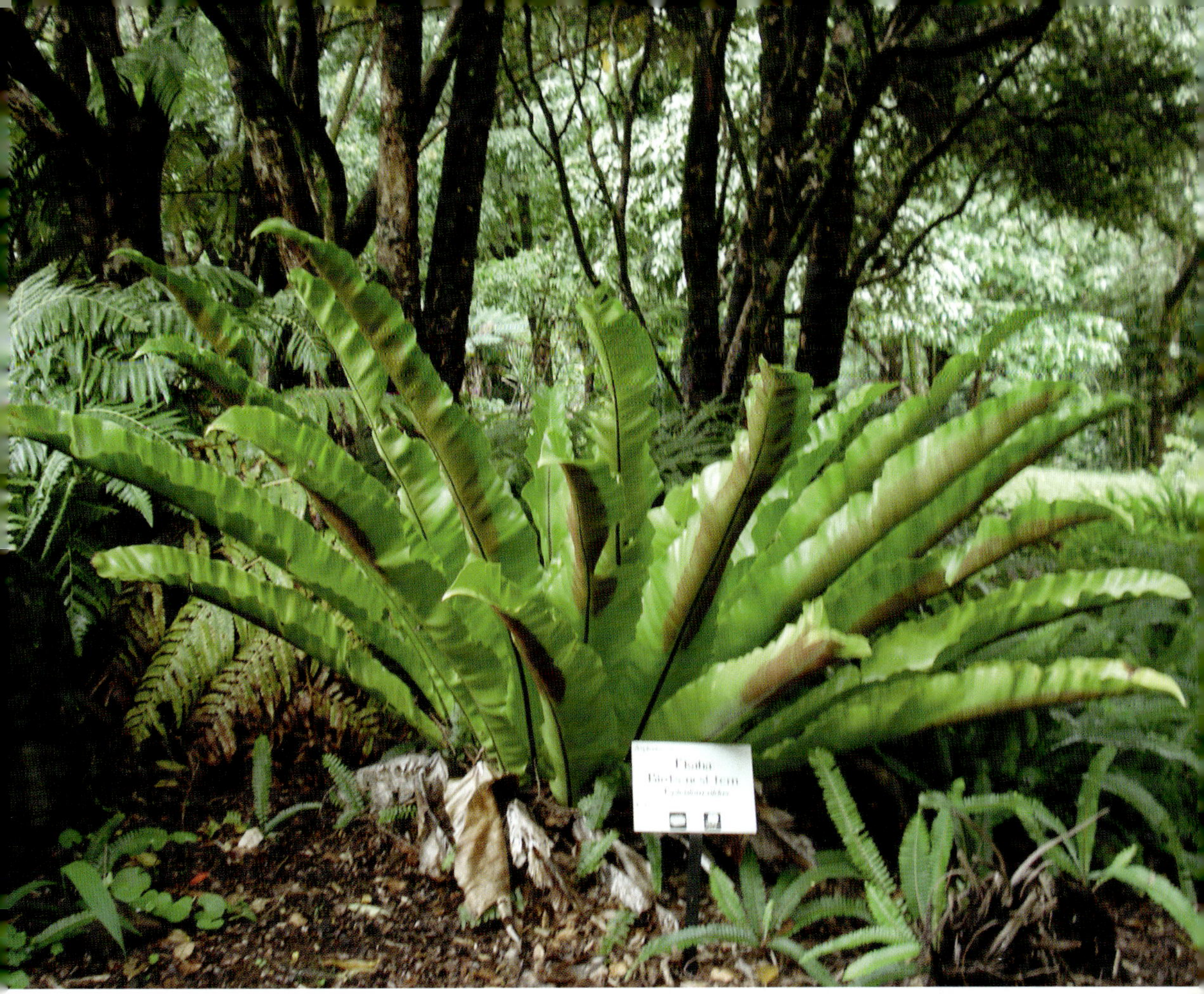

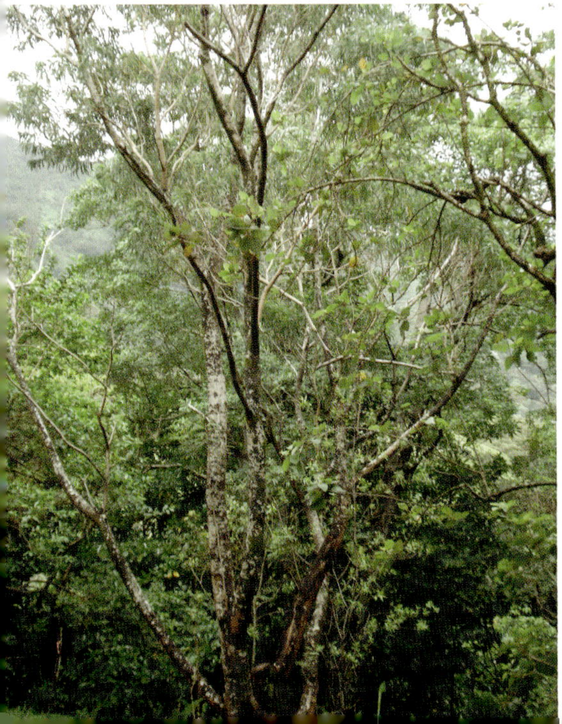

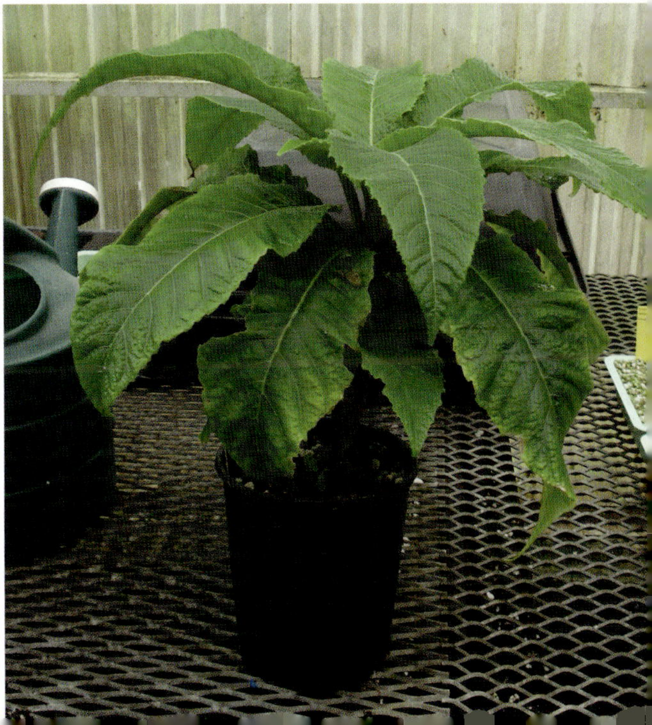

Local nurseries have also taken an active role in growing native Hawaiian species for use in landscaping and gardening. Coupled with a growing demand from the caring public, even national chain hardware stores in Hawai'i are purchasing natives from local growers for their customers' landscaping needs.

"Really it's about our sense of place," explains landscape architect Leland Miyano. "It's almost a moral decision to decide you want to grow native species because it's the right thing to do. I always like living pono, correctly."

Leland has been working on several native Hawaiian species projects. He is documenting the native species ho'awa (*Pittosporum confertiflorum*), trying to identify populations—specifically lower-elevation populations—of the genus on O'ahu. In addition to his documentation of the genus, Leland is propagating ho'awa throughout his one-acre garden at his home in the Kahalu'u Valley. He is also propagating and experimenting with 'ōhi'a trees, trying to see what types adapt best to a lowland environment. 'Ōhi'a trees are usually found at about an 80-foot elevation.

Loulu palm.

Loulu palms (*Pritchardia*) are the only palms endemic to the Hawaiian Islands.

"We cannot go to the mountains and collect native species and bring them here and expect them to grow," says Miyano. "All native species have their own ecological zone where they actually grew. A lot of the species that are available at the various outlets are species that are adapted to lowland coastal areas where the majority of our population lives anyway. Some of the ones available are the tough, low-maintenance plants, and if you can find the right plant for the right space, native species can be great."

Planting native species when possible shows a deep respect for the land and brings us closer as a whole to the natural beauty of Hawai'i. I doubt we'll ever see a loulu palm (*Pritchardia*) forest covering the 'Ewa plain as it once was, but we can plant loulu palms and other native species in and around our homes to stimulate the native environment. It is one more piece in the puzzle of *mālama ka 'āina*, care for the land.

In fact, the goal is not to uproot every alien species and replace it with a native species. On the contrary, the diversity of plant life in the Hawaiian Islands reflects the diversity of its human population. It is about achieving a balance between native and exotic, one that is beneficial to both parties.

Jill Laughlin, education and volunteer programs coordinator at the Lyon Arboretum in the Mānoa Valley on O'ahu, explains how exotics and natives can work together, "I think it would be difficult and probably an error to go in and completely start with a blank canvas, to remove everything and go down to bare earth and try to plant natives. You would have a situation where the plant scheme is so altered that it would be hard for any plant to get established in that environment where's there's no shade and only soil.

Native species micro-propagation lab, Lyon Arboretum.

Native Hawaiian hibiscus.

"At the arboretum we will use the canopy or the shade from an existing planting to establish native plants, and then we'll eventually drop the invasive tree and move it out of there."

For attracting butterflies Laughlin suggests planting māmaki, a hearty shrub with unusual white fruits and red-tinged veiny leaves that are a favorite snack of the Kamehameha butterfly. If you're interested in attracting birds, Hawaiian honeycreepers are attracted to red flowers like ko'oloa 'ula (*Abutilon menziesii*), a native hibiscus that almost fell into extinction, but is now known as an easy growing garden plant, suitable for lowland climates.

"It makes really important cultural connections," says Laughlin. "The plants that are here, the natives, really are the true representatives of Hawai'i and we need to honor that. We need to put them out there in the front. They have evolved in this habitat and have a definite right to be here. We've displaced them by our buildings and roads and all the activities that we do, and if we can somehow incorporate them back in, that's really important."

Here is a list of water-wise native Hawaiian species commonly used for landscaping:

BOTANICAL NAME	HAWAIIAN NAME
Ground Covers	
Dianella sandwicensis	ʻUkiʻuki (Lily family)
Heliotropium anomalum var. argenteum	Hinahina (Borage family)
Jaquemontia ovalifolium subsp. sandwicensis	Paʻuohiʻiaka (Morning Glory family)
Lipochaeta integrifolia	Nehe (Sunflower family)
Plumbago zeylanica	ʻIlieʻe (Plumbago family)
Sida fallax (creeping form)	ʻIlima papa (Hibiscus family)
Vitex rotendifolia	Pohinahina (Verbena family)
Small Shrubs	
Achyranthea splendens	ʻEwa Hinahina (Amaranth family)
Artemisia mauiensis	ʻĀhinahina (Sunflower family)
Wikstroemia uva-ursa	ʻĀkia (ʻĀkia family)
Medium Shrubs	
Abutilon menziesii	Koʻoloa ʻula (Hibiscus family)
Capparis sandwichiana	Maiapilo, Pilo (Caper family)
Gossypium tomentosum	Maʻo (Hibiscus family)
Hibiscus brackenridgei (state flower of Hawaiʻi)	Maʻo hau hele (Hibiscus family)
Nototrichium sandwicense	Kuluʻi (Amaranth family)
Osteomeles anthyllidifolia	ʻĒlei (Rose family)
Scaevola taccada	Naupaka kahakai (Naupaka family)
Sida fallax	ʻIlima (Hibiscus family)
Large Shrubs/Small Trees	
Chenopodium oahuensis	ʻĀheahea, ʻĀhea (Goosefoot family)
Dodonaea viscosa	ʻAʻaliʻi (Soapberry family)
Gardenia brighamii	Nānū, Nāʻū (Coffee family)
Hibiscus waimaea	Kokiʻo keʻokeʻo (Hibiscus family)

Large Shrubs/Small Trees (continued)

Myoporum sandwicensis	Naio, Bastard Sandalwood (Naio family)
Psydrax odorata	Alaheʻe (Coffee family)

Medium Trees

Erythrina sandwicensis	Wiliwili (Bean family)
Reynoldsia sandwicensis	ʻOhe (Ginseng family)
Sapindus saponaria	Manele, Aʻe (Soapberry family)

Vines

Canavalia galeata	ʻĀwikiwiki, Puakauhi (Pea family)

Sources: Fred D. Rauch and Paul R. Weissich, Plants for Tropical Landscapes: A Gardener's Guide (Honolulu, HI: University of Hawaiʻi Press, 2000); City and County of Honolulu, Board of Water Supply, "List of Plants at Halawa Xeriscape Garden" (www.hbws.org).

Xeriscape

There is a common misconception that xeriscape gardens are drought-resistant plantings, thorny cacti dispersed throughout a gravely bed. This might be the case in Las Vegas, but thankfully, we live in Hawai'i and therefore, we should plant according to our climate.

In actuality, xeriscape refers to a wise-water garden. It entails planting the proper plants that will thrive in a certain location based on the microclimate of that area. "If I can get away with just using rainfall, that's the best," says Miyano. Trees, shrubs, flowers, and ground covers are all part of a xeriscape garden and ideally, native plants that thrive on natural rainfall are your best choices.

The goal is to refrain from watering as much as possible and let the rain take care of that for you. If it rains a lot where you live, you can plant species that like a lot of water. If you live in a warmer and dryer microclimate, the water needs of the plants you choose should reflect that environment. If you're watering twice a day to keep a plant alive, it's probably not the right plant for your area.

Having a lawn is one of the biggest water consumers in the yard because grass is a very thirsty, fast-growing plant. If you must have turf, separate it from trees, shrubs, and other plants, so that they are not competing with each other for the same water source and can be irrigated separately. Better yet, replace turf with less water-demanding ground covers, plants, or mulches to coincide with the climate you live in.

Miyano is a proponent of getting rid of the lawn altogether and using low-maintenance ground covers instead. He explains how he uses the ground covers in his garden and their ecological benefit: "Have low ground covers that get tall enough to keep out the weeds and stay short enough to look nice. If you want to entertain, there are ground covers you can walk on

Rock garden.

The City and County of Honolulu Board of Water Supply has a demonstration xeriscape garden on their Hālawa grounds. The garden is open to the public and is set up to be an educational experience about alternative xeriscape gardening. Free tours are offered by appointment and self-guided tours and visits are available on Saturdays from 10 AM till 2 PM. Classes are also held on Saturdays; they hold an annual plant sale as well.

An estimated 50 percent of water consumption in the average single-family home is used outdoors. Xeriscape gardens can save anywhere between 30 to 80 percent in water consumption.

—Courtesy of HBWS

Right, A recycled brick path without cement creates water pathways.

and have a lawn-like surface, but don't have the water demand and the maintenance. I grow non-invasive ground covers that are easily gotten rid of if I have to take them out. I have a lot of ground covers planted in and amongst the natives, just waiting for the natives to grow, because a lot of the native species are a little slower to grow. As the native ground covers get a little larger and more vigorous, I take out the other ground covers. I grow herbs as ground covers too."

Here are a couple more ways to incorporate xeriscape principles into your garden:

1) Plan carefully. Incorporate meandering pathways, direct rainwater run-off around your garden. Place plants so that every type of plant gets the water it needs.

2) Mulch is a main component of a xeriscape garden in Hawai'i. It covers and cools the soil, minimizing evaporation, reducing weed growth, and slowing soil erosion. Mulch can be used around plants and trees or as an alternative to turf grass.

3) Good soil yields healthy plants. Improve your soil with soil amendments that have organic matter for better absorption of water and improved water-holding capacity. Roots need oxygen too, so aeration is also key.

4) If you must water, use a drip irrigation system rather than a sprinkler system. Drip systems put the water right at the base of the plant where it is needed most. Much less water is lost to evaporation this way.

5) Contact your local board of water supply for detailed xeriscape information and community support groups that can help you get started on your xeriscape garden.

—Courtesy of the HBWS

Natural Fertilizers

Along with energy efficiency, recycling, and other earth-friendly concepts and actions, people in Hawai'i are becoming more

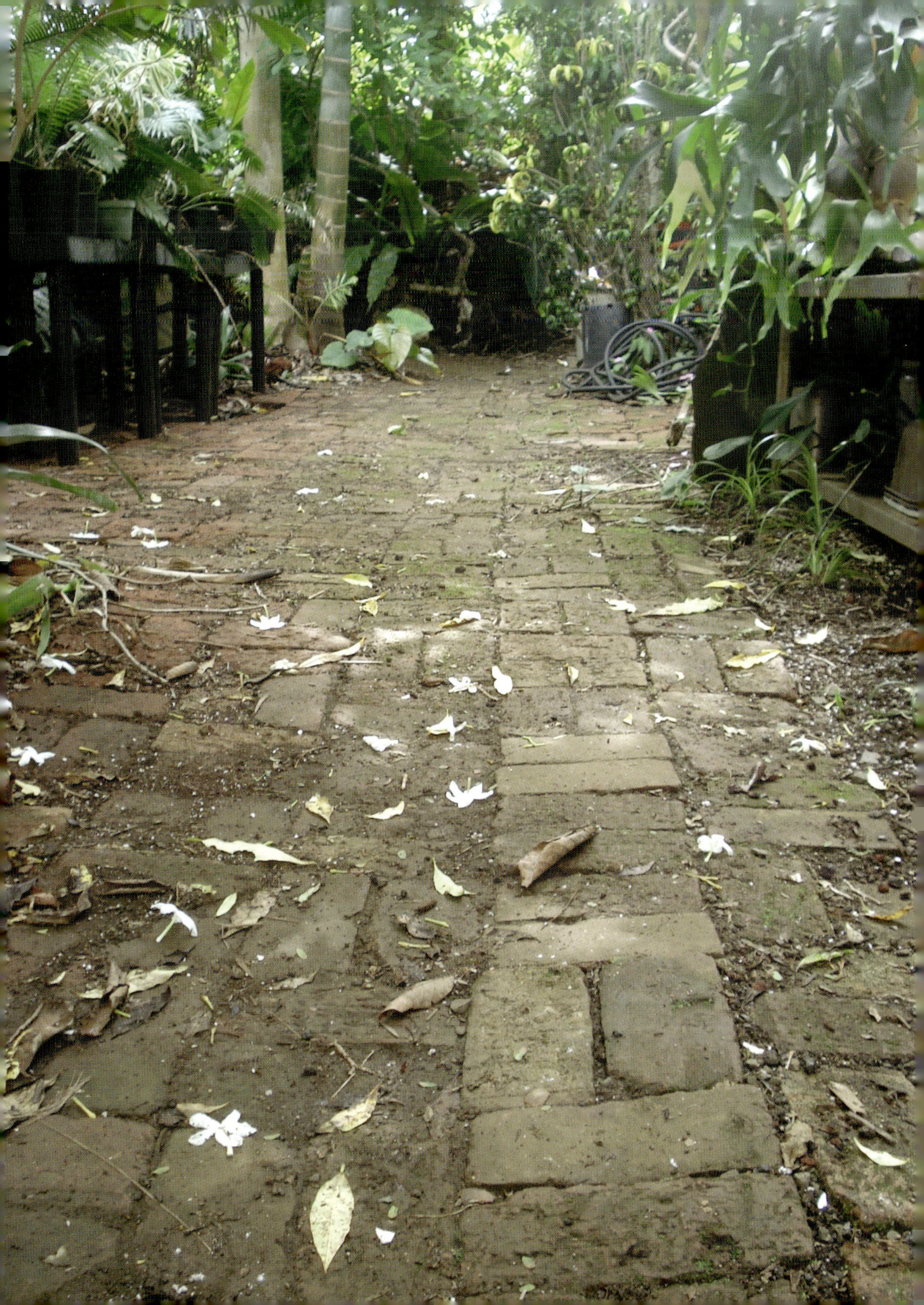

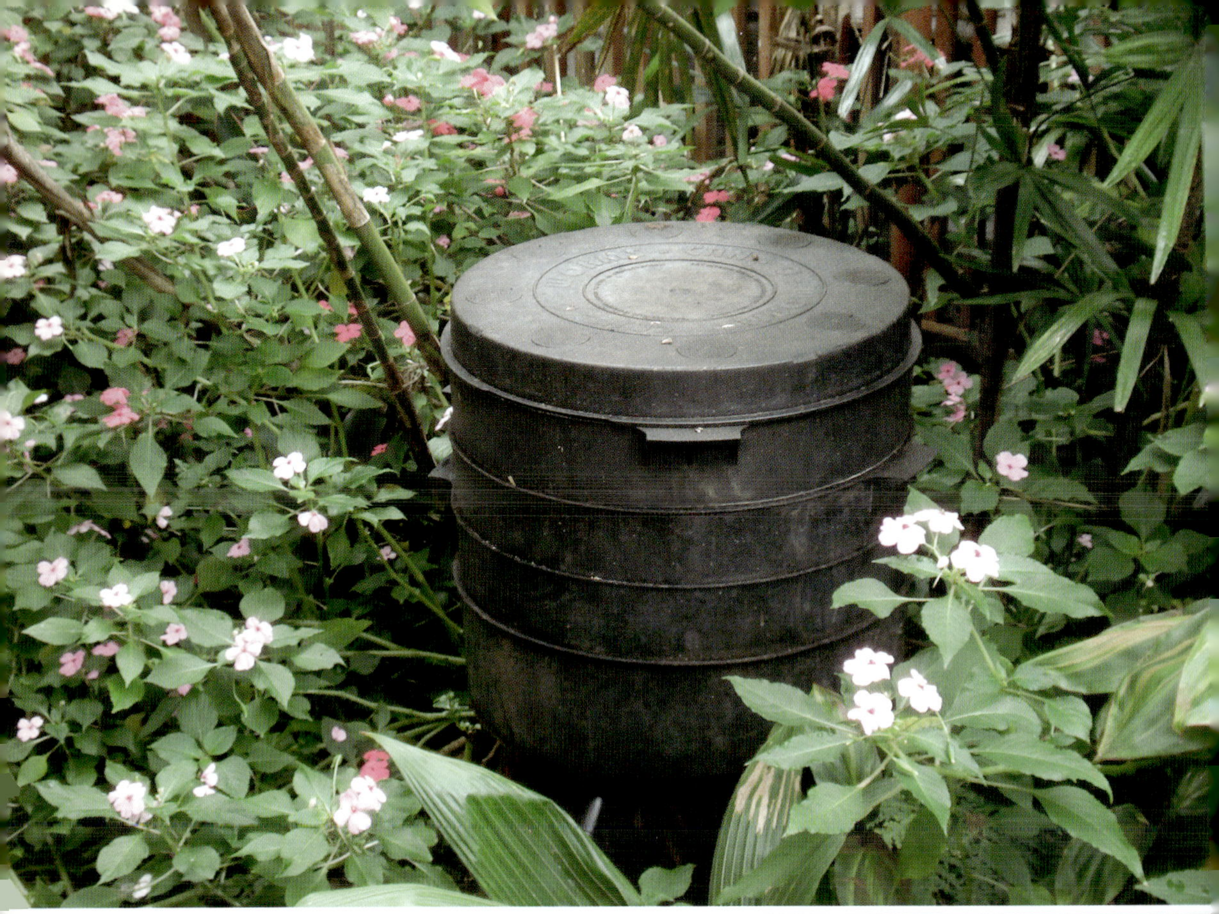

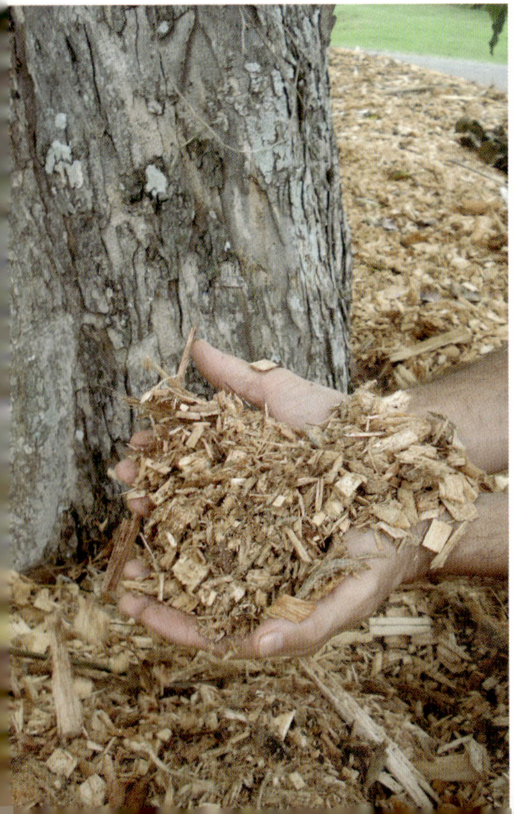

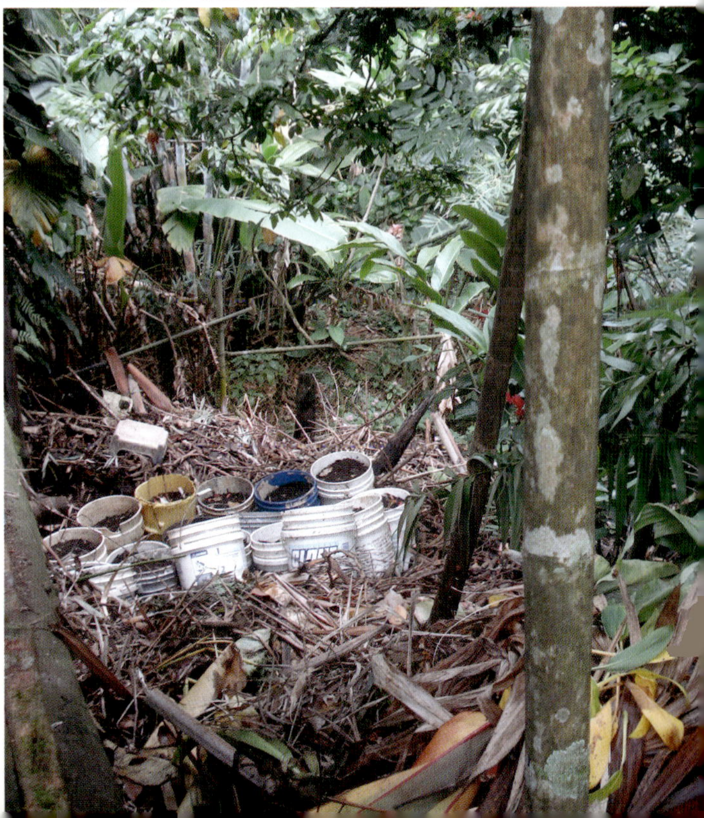

and more educated about and aware of what they consume, the corresponding waste, and the impact it has on the land and the ocean. While many people choose to use inorganic or chemical fertilizers to supply added nutrients to their landscaping, there are ways to fertilize naturally with organic materials. By taking kitchen waste and garden waste and composting or vermicomposting it, or using other green waste like wood chips for mulch, you can bring back an element of the natural ecosystem missing in most gardens and restore the natural nutrient cycle to your landscaping.

If you have ever been to one of the many beautiful and lush botanical gardens throughout the Islands, you will have noticed that many of the trees are encircled by what look like wood chips or possibly light-brown shredded leaves. That decaying ring of plant matter around the tree is mulch. It can be a variety of organic materials and its basic function is to promote plant growth and vigor by leaching nutrients into the soil and retaining moisture.

Mulch is naturally found in every rainforest as the layer of fallen leaves and branches that decomposes on the forest floor. Nature doesn't get things wrong, and we can use that example to benefit our own gardens and yards. Here are a few examples of mulch used in Hawai'i: grass clippings, cocoa shells, nut shells, tree bark, wood chips, and coarsely shredded tree leaves, twigs, and branches.

Here's how it works: the mulch is spread around the base of a tree or plant to a specific depth based on the type of plant and site conditions. The organic matter creates a nutrient-rich, porous barrier, which protects the bare soil from the sun, cooling the soil and aiding in water retention. If laid down properly, it also can inhibit weed growth. As the matter slowly decays, it attracts all types of beneficial bacteria, fungi, and insects that aid in the breakdown of the mulch and transform the organic matter into food for your garden. The nutrients make their way into the soil and plants consequently thrive. As you be-

> Red or black cinders can be used as inorganic mulch and placed around and throughout plant beds. Although the cinders will not add any nutrients to the soil like organic mulch, they still help the soil retain moisture and can keep out pesky weeds.

> Mulching is a great idea for reducing the need to water in drier areas, since the mulch helps the soil retain moisture. If you have very damp, soggy soil, do not use mulch as it will only retain the moisture and compound the problem. Increase drainage and amend the soil first.

> Some people put down plastic underneath mulch to suppress weed growth. Instead, try organic materials like banana leaves or fabric, like an old pair of jeans or a T-shirt.

come a mulching expert, the process will allow you to do away with chemical fertilizers, drastically reduce the time needed for weeding, and help you reach the ultimate goal: reducing the need to water the garden.

Nurseries and hardware stores sell different types of mulch prepackaged in large bags; some outlets will sell it by the yard for larger jobs. Check your local botanical garden to see if they have a surplus of mulch available to the public.

You can take an active role in making your own natural fertilizer through composting. Compost is similar to organic mulch in composition and function, except that it is farther along in the decomposition process before it is used in the garden; the finished product even looks like soil. It is a great way to recycle green waste into natural organic fertilizer and acts as an excellent soil amendment, providing aeration and nutrients to Hawai'i's soil, which tends to be very low in organic matter and therefore hard-packed and low in nutrients.

Take the banana peels, the coffee grinds, the lawn clippings, palm fronds, egg shells, plumeria leaves, and any other green waste you produce, including manures, and start a pile. A healthy compost pile needs oxygen, moisture, and heat. Insects and earthworms join bacteria and fungi in the task of breaking down the organic material into nutrient rich, stable humus.

There are two methods of composting: active and passive. Passive composting is very low maintenance—just let nature take its course and break down your pile of green waste. A much more efficient method of creating compost is through active composting, which entails turning the pile. When the pile is physically turned, it is aerated and the process of decomposition is rapidly accelerated.

Compost piles can be contained in a bin or box and can be small or large enough to correspond to the amount of green waste you produce. And no matter what the size of the compost pile, the benefits are the same. Used as mulch on top of the soil, compost helps retain moisture, control weeds, offset erosion, and of course, deliver valuable nutrients. When compost is tilled into the soil as a soil amendment, it improves drainage while retaining moisture, it increases bacteria and fungi populations in the earth that are necessary for nutrients to remain in the soil for absorption by roots, it improves soil fertility and aeration, and it restores a proper pH balance.

With a little research and some backyard experimentation, you can become an expert composter in no time and really cut down on the amount of waste that goes into the bin every week.

If you don't have the stomach to turn a compost pile and marvel in nature's magic power of decomposition, you can always let someone, or something, do the work for you. And there is one little critter that can't wait to help out: the worm.

Vermicomposting is the process of composting your kitchen waste and garbage with the help of worms. There are two types of worms used in Hawai'i for vermicomposting, the more common India Blue Worm (*Perionyx excavatus*) and the larger Alabama or Georgia Jumper (*Amynthas gracilis*). These worms are not "earth worms," the type of worms that are commonly found burrowing around in the soil. They are composters by nature and with the right requirements—dampness, darkness, and food—will thrive in a composting worm bin.

Worm-composting bins and a colony of worms should reflect the amount of kitchen waste you generate. Worm composting bins can be homemade or store-bought (there are mass-produced models on the market). Basically, a worm-composting bin has a bedding of damp newspaper, organic matter, worms, and some more damp newspaper covering the smorgasbord of scraps. The colony of worms will reproduce on their own in accordance with the amount of food they receive. If pickings are slim, the colony size will adjust to its environment. When the food is plentiful, the colony will multiply and grow.

The real benefit is the waste product of the worms that accumulates at the bottom of the bin—the worm castings. The

> Botanical gardens produce a great deal of mulch and compost. Check around with the staff to see if it is free for the taking.

worms turned your kitchen scraps into some of the best compost on the planet. Work the castings into your soil around indoor and outdoor plants for a healthy and fertile garden.

It is paramount that you buy locally grown worms when you start your colony. It is against the law to import worms from anywhere outside of the Islands. There is a $25,000 fine and the Department of Agriculture is serious about this law. Worms can carry in their gut fly larvae and other invasive organisms that cannot be detected upon inspection. The soil that worms are transported in is also a hotbed for invasive organisms, a free passage for harmful invasive species.

Interested in starting your own compost pile? Here are a few pointers:

1) Turn your pile every time you add new scraps to the pile. There are tools especially made for turning compost, or you can get creative with a shovel or rake. This way, oxygen gets through the pile to help break down the green waste.

2) Because decomposition of the compost piles relies on heat, the bigger the pile the better.

3) A compost pile needs to be moist at all times. This will depend on the items in the pile and the size of your compost pile, so you'll need to do your own monitoring and experimentation. Too much water and the pile will become soggy and stink. If the pile dries up, then decomposition comes to a halt.

4) Make sure to use a variety of materials in the pile. Too much of the same thing will upset the balance of the pile. Use kitchen scraps, lawn clippings, and yard waste in even amounts.

5) If you're composting in a bin, make sure there is adequate drainage at the bottom of the bin so your pile can get air from underneath and water can drain out.

Here are several local outlets that offer worm-composting bins of all sizes and the worms to fill them. Happy composting!

Worms in Hawai'i:
Waikiki Worm Company
234 Ohua Ave. #118
Honolulu, HI 96815
(808) 382-0432
waikikiworm@hawaii.rr.com
Mindy Jaffe, Owner
http://www.waikikiworm.com/1wwhome.html

Hawai'i Rainbow Worms
905 Hoolaulea St.
Hilo, HI 96720
(808) 937-2233
piper@hawaiirainbowworms.com
Piper Selden, Owner
http://www.hawaiirainbowworms.com/index.html

Joy of Worms
Maui
Wilma Nakamura, Owner
iris@wolbe.com

D. Kalua & Sons
Big Island
Darryl Kalua, Owner
(808) 982-6917
dmkalua@hawaiiantel.net

This link contains a guide to a do-it-yourself worm composting bin system: http://www.ctahr.hawaii.edu/oc/freepubs/pdf/HG-45.pdf

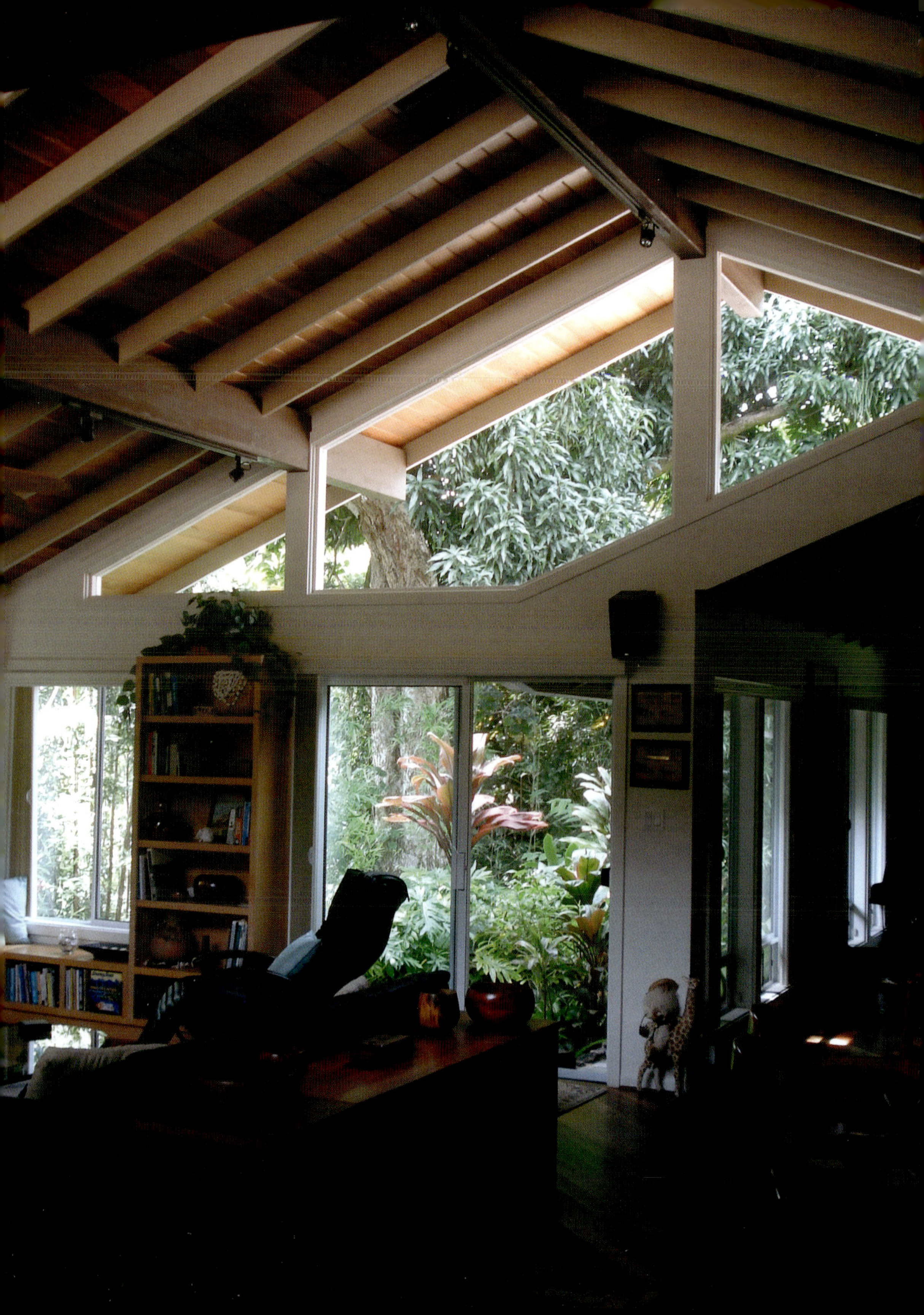

five: remodeling

There is no escaping the elements in Hawai'i. We are fortunate to have warm tropical sun, steady trade winds, and generous rainfall. These natural elements are also, however, the cause of most of the rapid wear and tear on our abodes, along with the help of pesky wood-hungry termites. Whether your home was built in 1954 or 1994, the signs of aging are no doubt visible.

A new roofline to increase natural lighting.

Owning a home is a significant investment and a 30-year loan is one of the only things in life we plan for that far in advance. To protect that investment, and over the years keep your home a comfortable and safe place to live, you will probably need to renovate your home, from the simplest thing, like some new furniture, to a complete makeover, let's say, a new roof.

Let's face it: replacing the roof is going to be expensive, but it is necessary. Call it an upgrade, or call it a makeover; either way, there is an abundance of greener materials, and contractors who specialize in energy-efficient and eco-friendly products and design.

The home-improvement market (and new home market) is responding to the demand for more recycled building matcrials and cn-ergy-efficient principles. Now it's up to you to choose the greener path to a greener home.

Floors

A wood or tile floor is just more practical in Hawai'i. It keeps your house (and pets) cooler and, unlike carpet, does not act as a fluffy home for mold and dirt. When it comes to wood flooring, some choices are more sustainable than others. Generally, look for a wood product that is harvested from a renewable source, like bamboo or eucalyptus. Both are fast-growing and can be farmed for their wood.

Here are some products on the market that offer greener solution to chopping down a forest for your kitchen floor:

Durapalm is wood from plantation-grown coconut palms from around the world. Coconut palms produce coconuts for about 80 years. Once they reach a non-productive stage they are farmed for their lumber and replaced with young coconut palms. The wood comes in a variety of colors.

Bamboo is one of the most widely used alternative hardwood flooring materials to date. It is available in many shades and different densities. Typically, strands of bamboo

---

Philip K. White & Associates in downtown Honolulu is known as a high-end residential and small commercial architectural firm, dedicated to designing homes and buildings with sustainable, environmentally sensitive, and culturally respectful practices. The firm doesn't just build green, they practice what they preach, as demonstrated by their offices Nu'uanu. The office, including the elevator, is powered mainly by a PV system. They have taken advantage of natural light throughout the office and incorporated an open-air roof deck—complete with a full bathroom—into the floor plan. All the faucets in the building are low-flow, and all of the carpet, adhesives, and paint are low in volatile organic compounds (VOC) for better air quality. Recycled and renewable materials like bamboo and eucalyptus were used, and all lights are equipped with occupancy sensors. Not only is the firm taking the eco-friendly high road, but they also provide a living, working example for clients in the community.

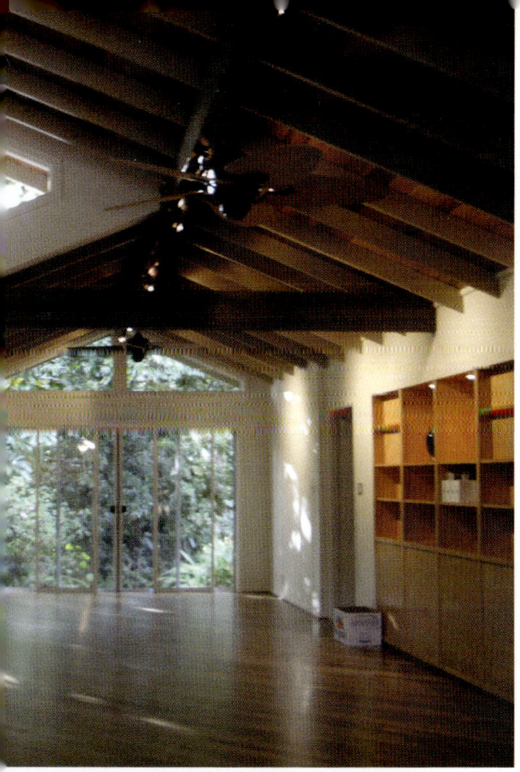
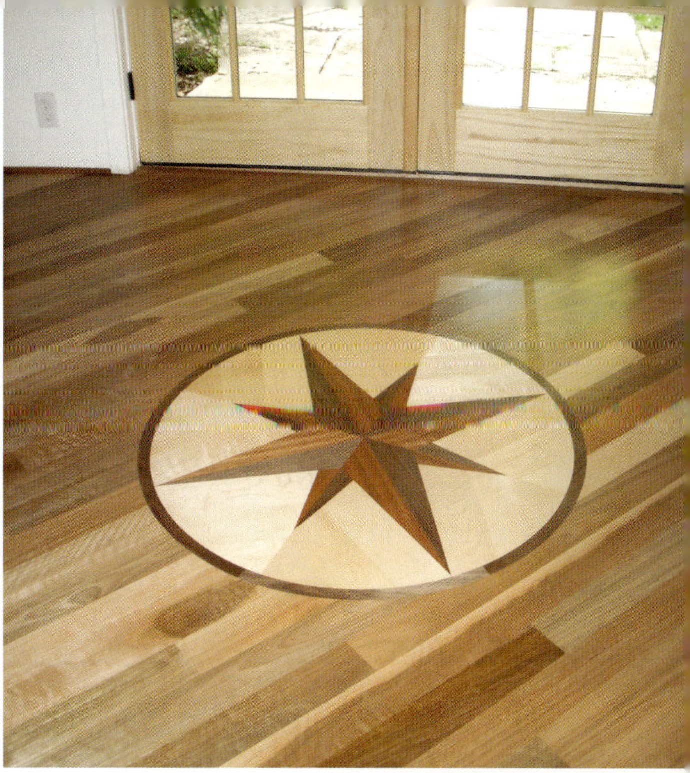
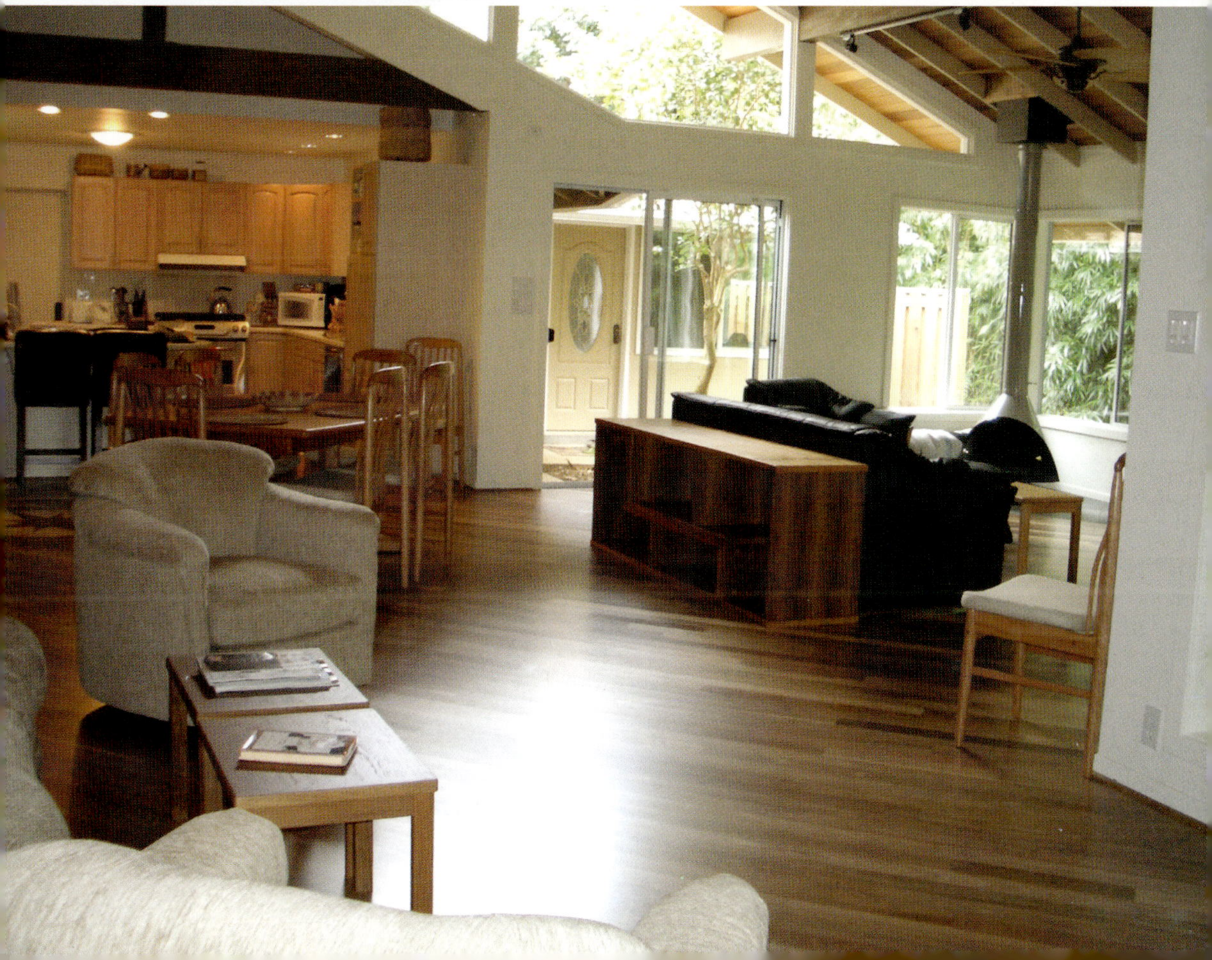

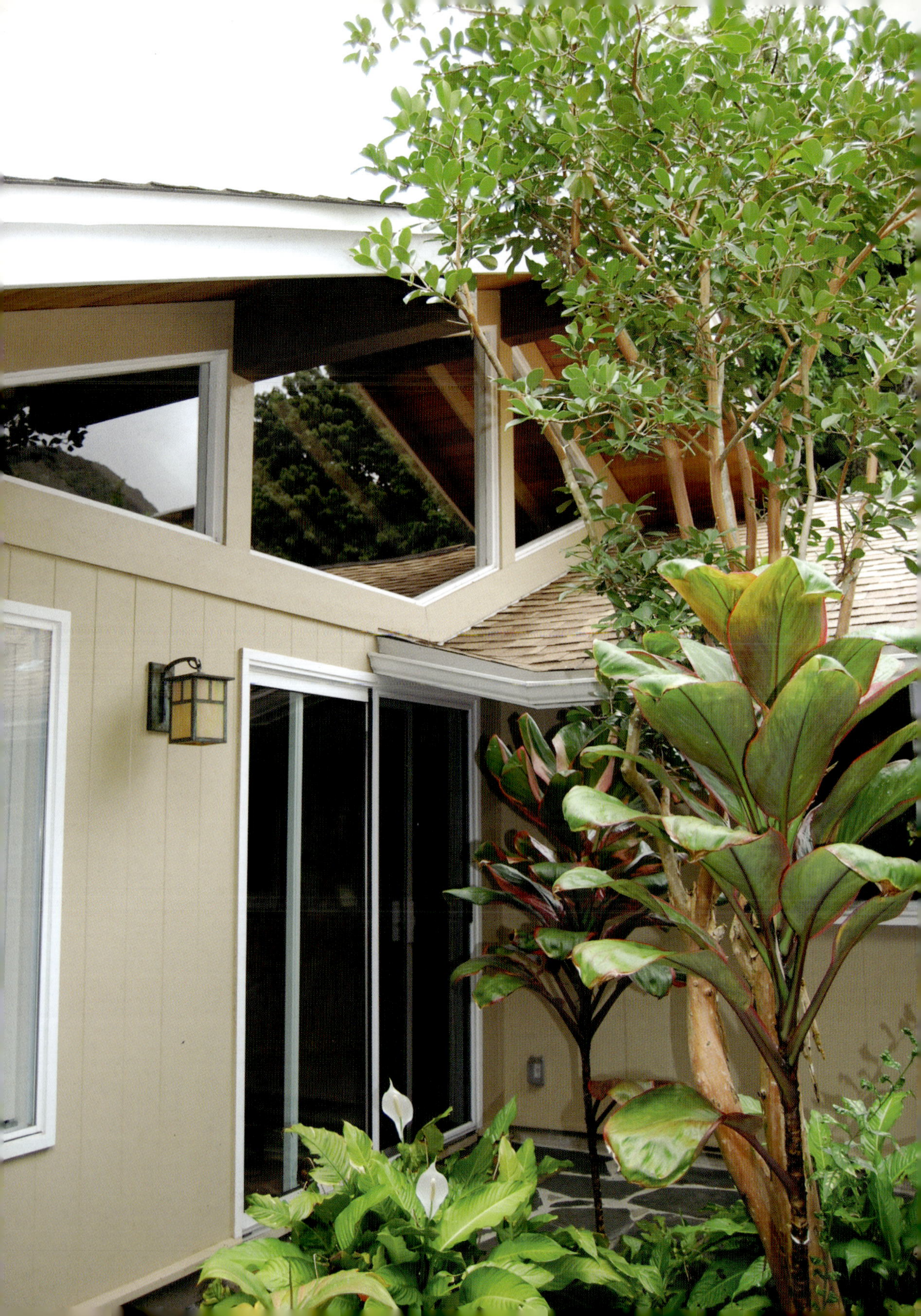

are shredded and compressed using a strong adhesive to form planks. Some bamboo planks can be compared to red oak or maple in hardness and density. Bamboo is more resistant to moisture than any type of wood—perfect for our humid environment. Ask your contractor or bamboo floor dealer about manufacturers who farm bamboo under an environmentally sustainable agricultural program.

Marmoleum is made primarily from natural raw materials such as linseed oil, rosins, and wood flour with natural jute backing. Its performance enhances over time as the organic product hardens and increases its durability. The oxidation of the linseed oil over the lifetime of the product gives the flooring anti-microbial properties.

Cork flooring is made from bark peeled from cork oak trees. The trees regenerate and thrive and are never cut down, so the material is completely sustainable. They are durable and known for their sound absorption and acoustical excellence.

Siding

When you consider the different materials for the exterior finish of your home, the key thing to keep in mind is resource-efficient siding, which is just a fancy way of saying siding made from recycled content. The goal in this case is to get away from using wood siding, which is easily weathered and prone to termite damage.

Metal sidings, such as aluminum and steel, are an option. They can vary in their percent of recycled content, and of course, the higher the percentage the better. The scrap is also recyclable.

Vinyl siding is inexpensive and very durable, but the PVC used to make the vinyl is very hard to recycle, so vinyl sidings include only a very small percentage of post-industrial scrap, if any. It is also low maintenance and withstands weathering much better than wood.

Re-use Hawai'i is a non profit organization that is dedicated to building-material reuse and recycling. They have great material for sale coming out of deconstruction projects and will also take home building materials that would likely be thrown away and added to the waste stream. www.reusehawaii.org

Light-colored sidings reflect light and keep walls cooler.

Radiant barriers can also be used in walls to promote cooling of the home by blocking the radiant heat from entering the home in the first place.

> A white or light-colored roof reflects sunlight and keeps the roof surface 20 to 40 degrees cooler.

There are also resource-efficient fiber-cement composites on the market. These composites are known for their durability, low maintenance, and good fire rating. Many also come with a 50-year warranty.

Roofs

The roofs of our houses take quite a beating from the elements. And there comes a time in most homeowners' lives when it becomes necessary to replace the roof. If you have found yourself in this situation, it's time to consider some of the many alternatives to the old resource-inefficient wood shingle.

For Hawai'i, our best bet is metal panels or composite shingles made with recycled content. Both materials provide low maintenance and high durability, so you won't have to worry about coconuts falling onto your roof and breaking off wood shingles.

There are several composite shingles, tiles, and panels worth a mention. Asphalt shingles are made with recycled waste paper and reclaimed mineral slag. There is also a roofing sheet material made from asphalt and fibers. Plastic shingles are made with recycled plastic resins and aluminum shingles, and panels can be made with up to 100% recycled content. Other composite roofing materials include fiber-reinforced cement products, which use recycled content and are very durable, but cannot themselves be recycled.

Kitchen with roof.

> Plastic lumber resists decay and insects, doesn't splinter, and is straight.

Plastic Lumber

Those two words don't seem to belong together at all. But lumber made from composites is one of the most durable, cost-effective, and low-maintenance ways to construct a deck, lanai, fence, or railing.

In Hawai'i, wood has to be one of the most inefficient building materials due to our humid and damp climate, combined with wood's tendency to warp and rot. Wood has to be sealed, treated, and maintained to prolong its life or it will most likely fall victim to termites. Not to mention the resource is finite, and the regeneration of farmed trees is a long and slow process, even if they are fast-growing species.

No sense cutting down trees when we can buy lumber made from recycled plastics. Some types of composite lumber are made entirely of recycled plastic content; others are made from a combination of reclaimed wood and plastics, like recycled plastic trash bags and waste wood fibers. Composite woods come in different colors that mimic different types of real wood and also different cuts for specific applications in the design.

As in all things in life, there are many perspectives to take into consideration. Some schools of green thought say that wood is our best building resource in the Hawaiian Islands. They claim it is a renewable resource and requires less consumption of fossil fuels to process than the milling and subsequent recycling of steel. As a consumer, it is always a good idea to educate yourself on all sides of the issue or product so you can make the best choice for yourself, your family, and your lifestyle.

> Plastic lumber is versatile in form and function. When using it in structural applications, be sure to use high-grade plastic lumber. Here are a few uses where plastic lumber can be used: marine docks, decks, fencing and posts, outdoor benches and chairs, boardwalks, boat trailer runners, playground equipment, landscape timber, planter boxes, and retaining walls.

Roof line appears.

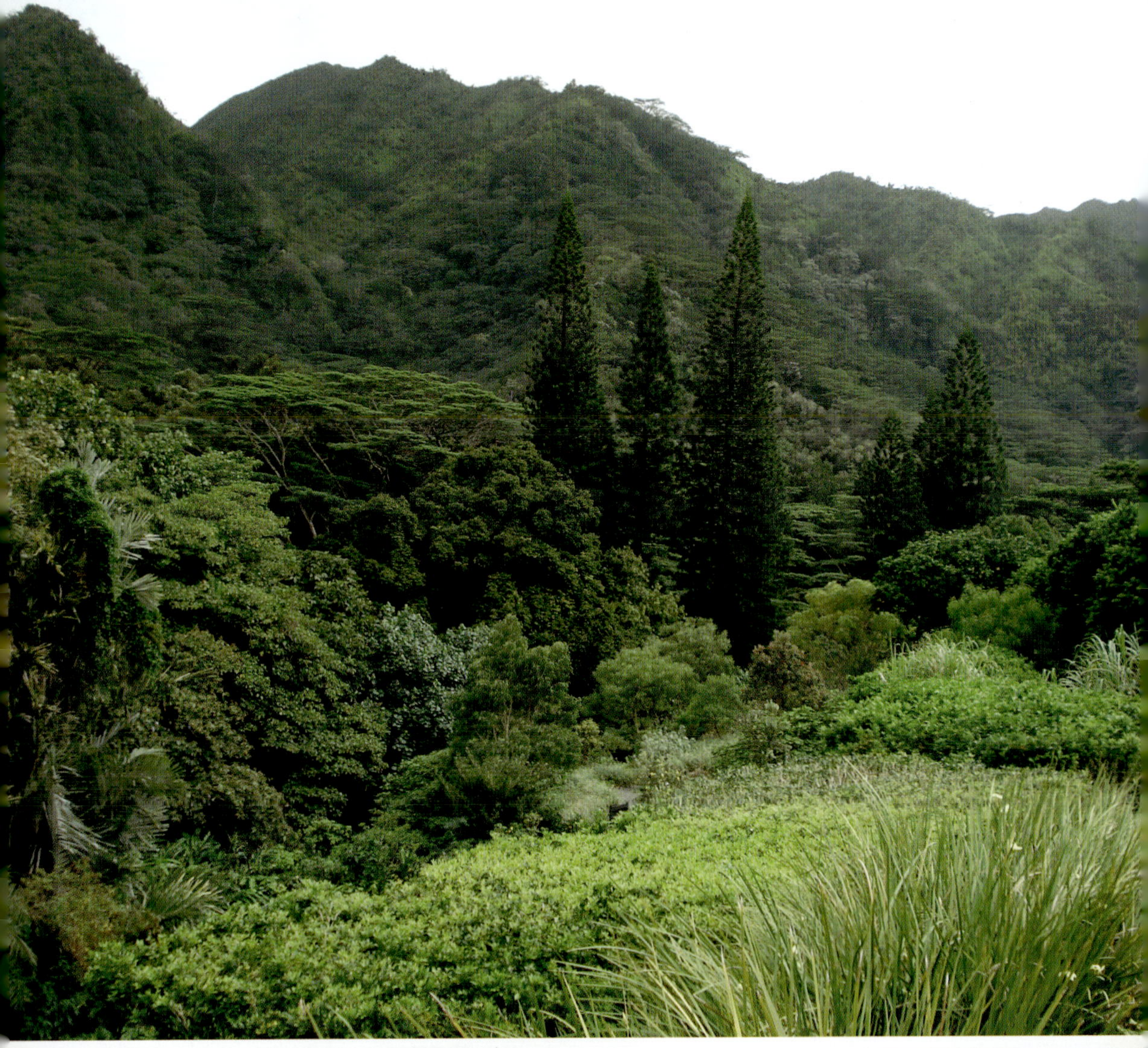

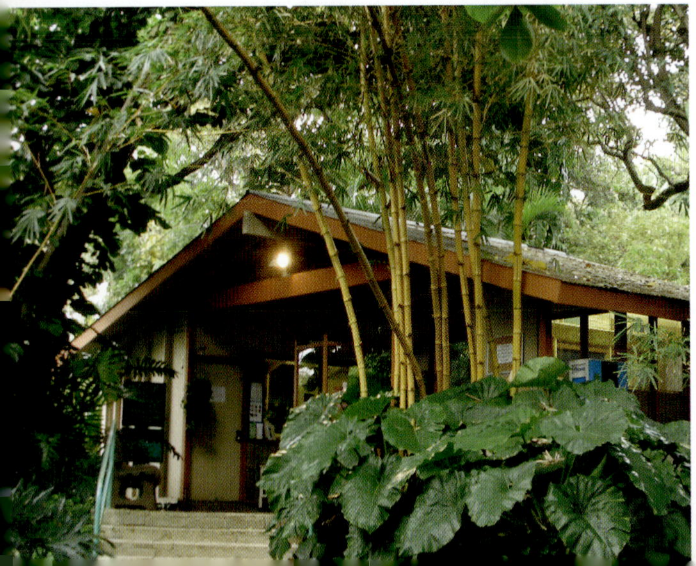

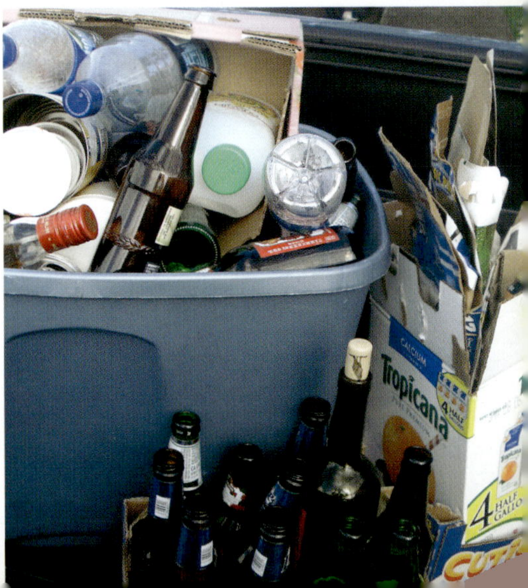
Tropicana

IN THE COMMUNITY

part 3

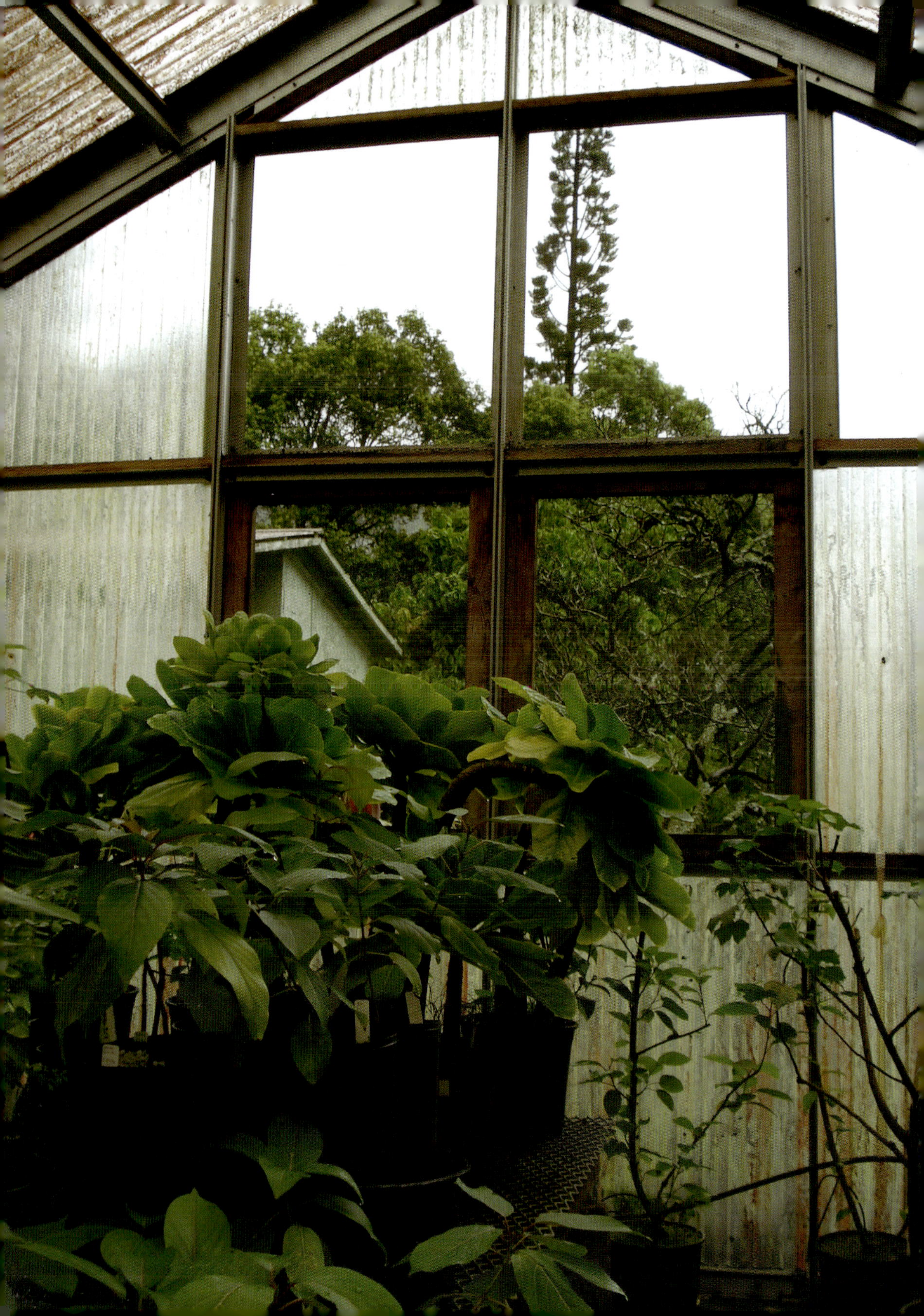

six: consuming

When we look at energy efficiency and sustainability in our homes, it is the little things that we do every day that add up to achieve our overall goal. Each change toward sustainability for the individual is a step forward, a step in the right direction. Combine your energy savings with everyone on the block and all of a sudden that one small action, using CFL bulbs instead of incandescent bulbs, is multiplied throughout the community.

Native species greenhouse, Lyon Arboretum.

And that is why, as we move beyond the walls of our home, our goal of energy efficiency and sustainability does not end. In fact, it extends down every dirt road, through every housing tract, across the intersection and into the malls, restaurants, and grocery stores on every island. Big city, small town—no matter.

> Take the ecological footprint quiz at www.earthday.net/footprint and find out how much you are actually consuming. Better sit down for this one.

As consumers, our power lies in our purchasing power, in our decisions to support local farmers and storeowners, to buy organic products, and to buy or reuse recycled goods. When we buy certain products, we let small businesses, corporations, and manufacturers know exactly what it is we want. And they will listen if they want to make a profit.

And with our power as consumers comes another important responsibility toward our goal of a green Hawai'i—consuming less of everything. The less we consume, the less is required to produce those goods. and the less that goes into the landfill. By avoiding products with too much packaging or those made with materials that are not recyclable, we can send a strong message without spending a dime.

So let's continue our journey toward a greener Hawaiian home and check out a few ways we can use our almighty dollar in an efficient and sustainable manner.

Buy local goods

Just about everything we consume in Hawai'i, from the rice we eat to the paint on our house, is shipped in from overseas. We are, more or less, dependent on the container ships and the goods they deliver. And whether from the mainland or China, the goods carry with them a substantial carbon footprint.

Consider the fuel and energy needed to process or manufacture a product, to package it, to transport it to the docks and load it into a container, to ship it a few thousand miles across the ocean, to unload it and transport it to a store near year. A head of lettuce that makes that journey from the mainland uses a lot more energy and natural resources and is responsible for much more pollution than a head of lettuce grown in the Islands. That is why buying local is so very important to a greener home.

When you buy local you support local farms, families, businesses, communities, and industries. The energy and natural resources saved in your decision to buy local are no different than using a CFL bulb instead of an incandescent bulb. The goal is the same: to reduce our dependence on and use of inefficient fossil fuels for energy.

Farmers markets are great places to find fresh produce, live herbs, orchids, other landscaping plants, and even a tasty plate lunch. By shopping at farmers markets, we are supporting the local farmers who grow the produce and in turn, keep our community strong and prosperous. You can often find crafts, garments, jewelry, and art at farmers markets as well.

If you live on Oʻahu, Chinatown is a great place to find home-grown produce. Talk to the employees or owners and find out if their produce is indeed locally grown.

The list goes on and on. From a great selection of locally crafted beers to local Maui Rum, there are all sorts of interesting local products to discover and enjoy.

> Check out www.thestoryofstuff.com for an interactive, in-depth look at where all our stuff comes from, where it goes, and how it impacts our lives and the environment.

Slow Food

There are two basic types of goods: things we want and things we need. Food falls into the latter category. But for most of us, the days of subsistence farming are long gone, and now we spend our days working for a paycheck in an office, not in the fields. That means we spend a lot of money purchasing the food we consume.

Buying food is a great example of how your purchasing power can make a difference and how important it is to buy local. With fast food available on almost every corner, and shopping malls full of chain restaurants, a movement was initiated to bring the focus back to local foods and stress the importance of a good, home-cooked meal.

Slow Food is an international organization of foodies, dedicated to eating local, seasonal, and organic foods, and enriching their quality of life by slowing down and enjoying tasty, wholesome meals with friends and family. The grassroots organization has quite a foothold on the Big Island, but is still in a fledging state on Oʻahu. Slow Food members are a great example of in the community • 90

Check out www.slowfoodhawaii.org or www.slowfoodoahu.org for more info. The Oʻahu website has a list of some farmers markets and links to local farms, and the Big Island website offers a list of markets, farms, and restaurants to get you started on the road to eating and buying local.

A head start for these native species: Lyon Arboretum native species greenhouse.

how choosing where to spend your money can make a difference in the community. When we support local farmers, the money stays in the community, bolstering local agriculture and the economy.

Karen Miyano, Slow Food member, explains the importance and ideals of the organization: "It's about sourcing out local foods, traditional foods, and growing your own if you can. But it's more than that. It's about coming back to the table with family and friends and engaging in the preparation of foods—that whole dynamic of being around a table, sharing food, making food. That time seems to be so lost in people's lives, eating on the run or sitting in front of the TV with food and not engaging in conversation."

This issue is important to Miyano and others simply because Hawaiʻi has the ability to be more self-sufficient, yet we rely so heavily on imports. She adds, "People need to go to the grocery stores and ask, 'Is this locally grown?' and request that they have locally grown products because they have to listen to the consumer. The consumer has power because [the retailers] are completely relying on you.

"We are an isolated place and so reliant on imported food. When the ships aren't coming, if there's a strike, it could be devastating. There is about an 11- to 15-day food supply on the island, it's scary. You can't live on rice and water. But we have the resources in the Islands to feed ourselves."

Green Cleaning Products

Air quality is an important part of a green home because it adds to your comfort and quality of life, and is paramount to creating a healthy home. So when it comes time to keep your home clean, consider using eco-friendly cleaners instead of harsh chemical detergents.

Have you ever been using a chemical cleaner, say, to clean the shower, and the smell was overwhelming to the point of irritating your nose and making you cough? Ammonias and bleaches are harsh and can be dangerous when used improperly or in large quantities.

Eco-friendly cleaners are usually citrus-based and have a pleasant smell, and can still get the job done. And with the demand for alternatives to harsh chemical cleaners, there are more and more choices popping up everyday for an assortment of applications.

If environmentally safe cleaners can clean your bathtub, then they can, no doubt, clean you as well. Try castile soaps and shampoos when you shower instead of their harsher soapy counterparts made from detergents.

Simple Green has cornered the market on environmentally friendly cleaners for more than 30 years. Their products are non-toxic, biodegradable, non-hazardous, non-flammable, and non-corrosive. You can wash the car in the street with Simple Green and the soapy runoff water that heads to the ocean will not pollute or harm anything downstream. And it's concentrated, so a little goes a long way.

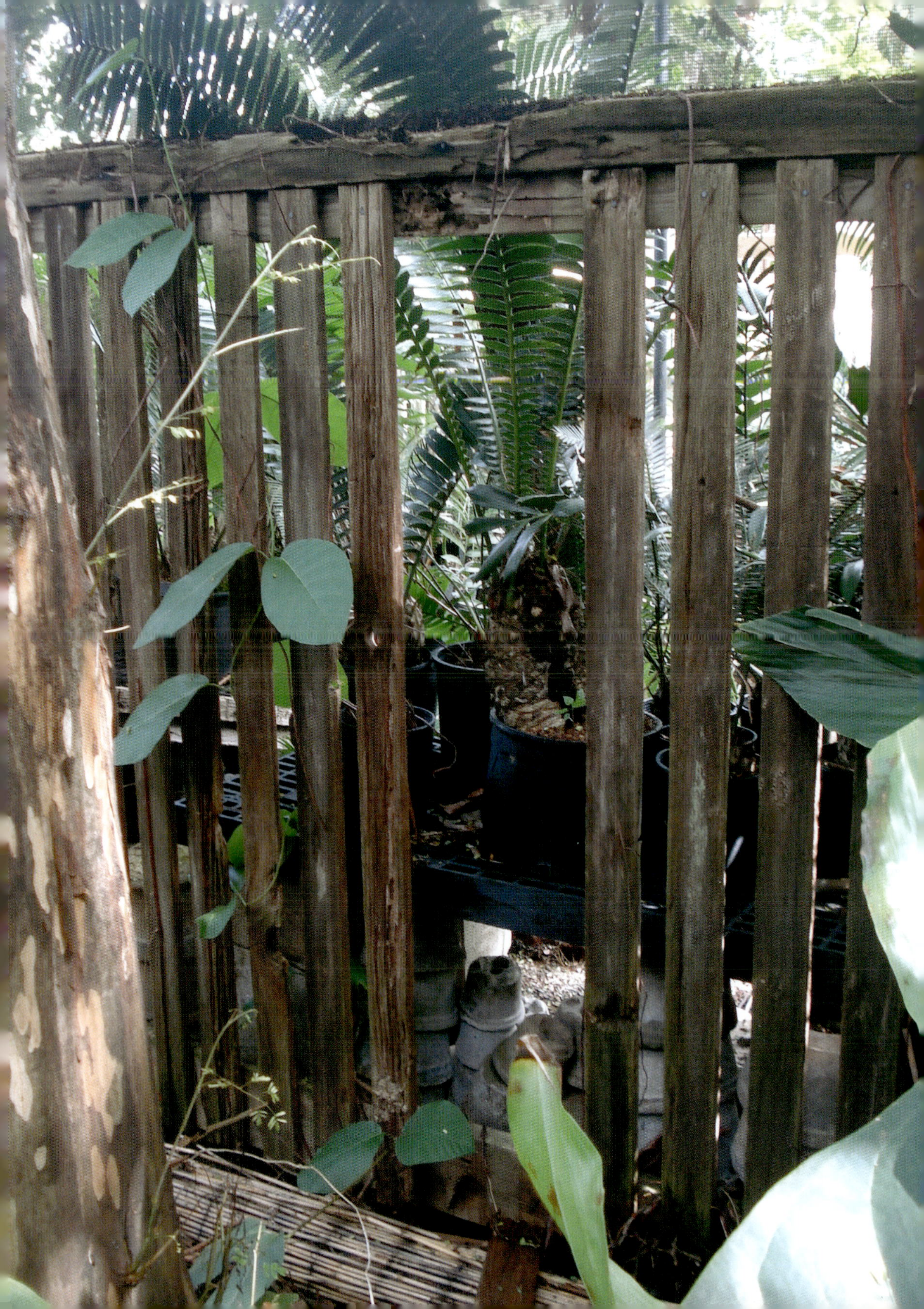

seven: recycling

Recycling is a strategy that is ingrained into the human psyche, a phenomenon that has occurred throughout human and Hawaiian history. Of course, in ancient Hawai'i there were no plastic bags, no paper boxes, and certainly no aluminum cans. But the fundamentals of recycling—reusing a material to make something else—occurred nonetheless. Hawaiians wove human hair and used whale-ivory for *lei niho palaoa*, ornamental necklaces worn by the ali'i as a symbol of rank and authority; they made fishhooks from wood, human bone, shell, and turtle shell; they used *hala* (pandanus) leaves for thatching and weaving into mats; they used kukui nuts for oil and suit; they used bird feathers for decoration; and they made tools from stone, shells, and coral. All of these are forms of recycling.

Restaurant sink salvaged and reused in the yard.

We think of recycling somewhat differently today, as our materials and lifestyles have changed significantly. But recycling still happens on all levels. Whether you take aluminum cans to the recycling center for your deposit, or you salvage old bricks to make a path through your garden, the principle is the same: reduce, reuse, and recycle.

In ancient Hawai'i, as in other places in the world, even in modern times, recycling or reusing materials was necessary for survival. But with our first-world throw-away society of consumerism, plastics, glass, and metals can be recycled and reused, reducing the amount of trash and extending the longevity of our landfills.

Recycling, HI-5.

HI-5

According to the Hawai'i State Department of Health, approximately 900 million beverage containers are sold in Hawai'i each year. Can you imagine all those recyclable containers taking up precious space in landfills? Luckily we don't have to, because in 2004, the Hawai'i State Legislature voted into law the Deposit Beverage Container Program, also called the Bottle Bill. In November 2004, the state required eligible beverage containers to be marked with the "HI-5" label. Consumers are charged an additional six cents for each beverage container purchased: one cent is a non-refundable fee that pays redemption centers to process and recycle the containers, and the other five cents are a refundable deposit returned to the consumer when they take their empty containers to the redemption center to be recycled. At the redemption centers, small amounts of beverage containers are counted and larger amounts are weighed to determine a cash refund amount.

The program has been a huge success since refunds began being doled out on January 1, 2005, and annual redemption rates have climbed every year since. Between July 1, 2006, and

June 30, 2007, more than 636 million containers were redeemed statewide, which translates to a 68% redemption rate.

When you return your recyclables to a redemption center, your items are weighed and a refund is based on a standard conversion rate. You can request hand counts of up to 200 beverage containers, ensuring an extremely accurate return. Recently, the State has added a small plastic bottle rate for plastic bottles of 17 fluid ounces or less, but you have to separate those yourself from the rest of your recyclables if you want to take advantage of the new rate.

These incentives were designed to encourage consumers to recycle their beverage containers, and the program is paying off. So let's continue to keep the cans and bottles off of the side of the road, out of the streams, and off the beaches, and instead take them to a recycling bin.

The HI-5 program website, www.hi5deposit.com, is an excellent resource for almost all of your recycling needs, from community and school programs to all the facts, figures, and information you need to get started.

Redemption Centers

Even though the HI-5 beverage container deposit program is run by the State of Hawai'i, each county oversees local recycling efforts. In turn, the redemption and recycling centers are

Community recycling bin.

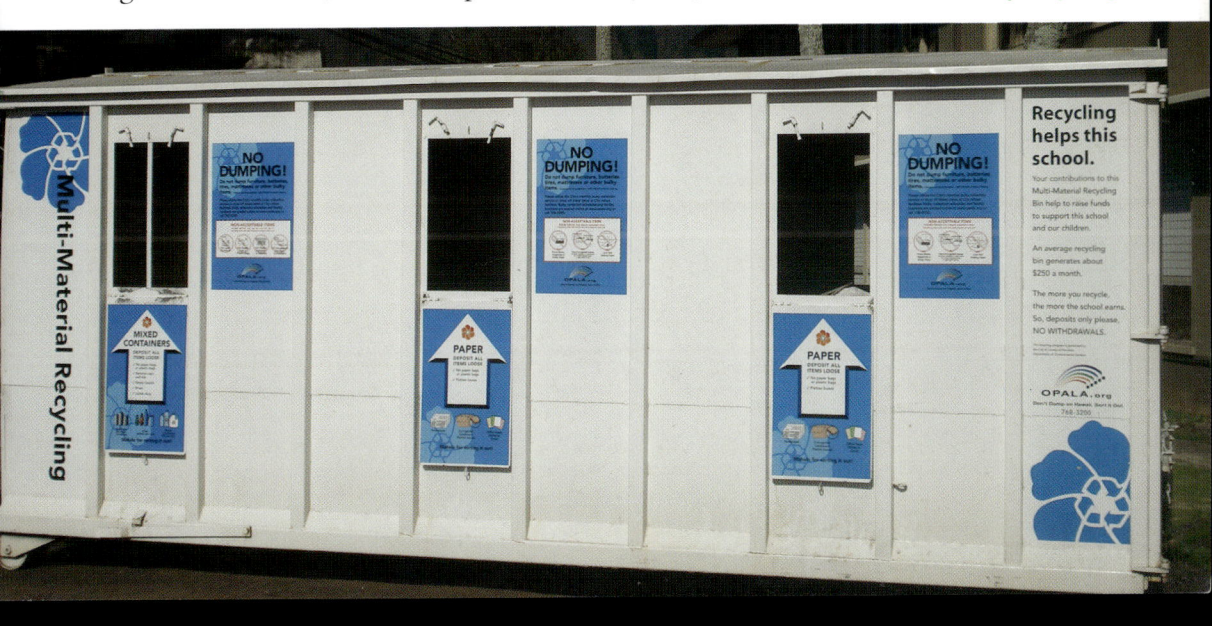

Plastics coded #1 and #2 are recyclable in Hawai'i.

Beverage Container Types

All non-alcoholic drinks: soda, water, juice, tea, coffee, except for milk or dairy products.

Limited alcoholic drinks: beer, malt beverages, mixed spirits, mixed wine.

Container material type: aluminum/metal, glass, plastic.

For a complete list of what containers are accepted go to www.hi5deposit.com/support/ConsumerInfo.pdf

Plastic Containers

Only plastics #1 and #2 are covered under the Bottle Bill. Look for the triangle and number recycling symbol on the container.

On December 1, 2007, two-liter containers were included in the HI-5 program. A two-liter container is defined as 64 to 68 ounces, and the same deposit and refund rates apply. As of publishing, the law applies to containers 68 fluid ounces or less that are marked with the HI-5 label.

Containers must be empty and clean. Remove the cap and do not crush the container.

To Count or Weigh: Redemption Rates

One pound of aluminum products = 31.6 containers
One pound of glass = 2.3 bottles
One pound of plastic bottles = 17.5 containers
One pound of bi-metal products = 8 containers
One pound of small plastic bottles (17 fluid ounces or less) = 22.7 containers

private businesses. Check out a list of redemption centers on your island at www.hi5deposit.com/redcenters.html. Since redemption centers are popping up all the time with the increasing interest in recycling, and redemption centers can be mobile, dial 2-1-1, the HI-5 hotline, to keep current.

The amount of things to recycle is almost endless—from Christmas trees to motor oil, you name it—and there is a wealth of recycling information to explore for residents on all islands.

Recycling is managed separately on each island by the government at the county level, so check with your local county to find out what's going on in your community and all the different ways and things you can recycle:

City and County of Honolulu, Department of Environmental Services, http://envhonolulu.org/solid_waste

Maui County, Department of Environmental Management, www.co.maui.hi.us/departments/EnvironmentalMgt/Recycle/index.htm, Recycle Maui County Hotline: 808-270-7880

Kaua'i County, Division of Public Works Recycling Programs, www.kauai.gov/Government/Departments/PublicWorks/RecyclingPrograms/tabid/68/Default.aspx

County of Hawai'i, Department of Environmental Management, www.hawaii-county.com/directory/dir_envmng.htm

Donating Your Recyclables

When you donate your recyclables to community bins found in school parking lots, you can give yourself two pats on the back: one for recycling and the other for donating to your local school. Those large, white, 40-cubic-yard community recycling bins accept all the HI-5 beverage containers, wine bottles, glass bottles and jars, all plastics with code #1 or #2 embossed on the container, newspapers, corrugated cardboard, and white and colored office paper. Cereal boxes, junk mail and magazines, plastic bags, tuna cans, and plastics coded #3 through #7 all go into the trash where they are sorted at refuse facilities. Your good deeds will help ensure that proceeds from the sale of the materials go to the schools and right back into the community.

If you don't have a multi-material white community bin at your local school, contact your county recycling coordinator about getting a bin for your school. There are also HI-5 community collection bins for the donations of HI-5 beverage containers only. A full HI-5 collection bin can net an estimated

---

Recycling one aluminum can saves enough energy to run a TV for three hours.

—Courtesy of the State of Hawai'i Department of Health

Check out www.recycle-hawaii.org for information, news, and events related to recycling on the Big Island.

In addition to HI-5 recyclables, community non-profit organizations like the National Kidney Foundation, the United Cerebral Palsy Foundation of Hawai'i, and Big Brothers Big Sisters of Honolulu accept all sorts of recyclable donations like cell phones, small household items, used clothing, and vehicles.

But what about furniture, paint, or storage containers? You can get all these things and more made from re-cycled content at www.recyclestore.com.

$1,200 for an individual school. You can also check out other school fundraising programs at http://envhonolulu.org/solid_waste/Recycling_Fundraisers.htm.

Follow this link, www.hi5deposit.com/schoolrecycling.html, for community contact information and a guide to fundraising through recycling for schools.

Reuse and Reduce

Now that you've taken your containers to the community bin to be recycled, there is still one more loop in the triangle of recycling: reuse. When we reuse products instead of buying new ones, we are consuming less. The harvesting and processing of raw materials slows, and the need for excess packaging is reduced. Environmental impacts related to shipping are eased and our generated waste decreases. Often, the choice can be as simple as spending a little more money for a durable and long-lasting product over a cheaper and flimsier one that would need to be replaced more often.

The ability to reuse things is endless, limited only by your imagination. Building materials such as lumber, bricks, and rocks can be reused in the garden to build an orchid house or a greenhouse; for shelving for plants; and for creating pathways, dividers, and edging. Jeans and cotton shirts can be spread under a planting bed instead of black plastic to prevent weed growth. Old tree stumps work well as plant stands. You can use a leather work boot for a plant container, and dresser drawers can be dressed up and turned into artsy, wall-mounted shelves in the house. If nothing else, many things, like old surfboards or skateboards, can be used again as a canvas for art.

When we make a concerted effort to reuse what we have to the fullest potential of the object's lifespan, we eliminate the need to buy new things and keep excess trash out of the landfills. By reusing goods whenever possible, we effectively reduce our impact on the environment, from the need to cut down more trees to the problem of overflowing landfills.

Recycled Products

There is also another way we can reuse products, like plastics, that have already been manufactured and produced. Once materials have been recycled, they can essentially be reused to make new things. It's as if they return into the manufacturing cycle as raw materials, but without taking more untapped natural resources from our Earth.

Products made from recycled content are a staple in the production of eco-friendly goods, and their popularity is growing. Plastics are being used to produce T-shirts, hats, shopping bags, and other goods. Car and bicycle tires are recycled as durable soles for shoes. Check out www.clothesmadefromscrap.com for clothes made from recycled plastics and reclaimed cotton.

Hawai'i's connection with surfing and world-class waves has introduced a new forum for recycled goods. Surf apparel companies such as Billabong and Patagonia are offering recycled-content boardshorts for environmentally conscious water enthusiasts.

Buying recycled furniture is another great way to take advantage of recycled or reclaimed materials in your home. Recycled furniture doesn't mean pulling someone's old sofa out of the dumpster. There are retail outlets all over Hawai'i that carry beautiful wood and fabric furniture fashioned from recycled materials. The only way you can tell the furniture is made from recycled product is from the tag that boasts of its eco-friendly production and material.

Two great examples of beautiful recycled furniture are reclaimed teak furniture, available at Bali Moon (www.balimoon-hawaii.com) and outdoor furniture made by furniture maker Janus et Cie, constructed of nontoxic and recyclable fiber and available at O'ahu retailer Pacific Home (www.pacific-home.com). The reclaimed teak furniture is new, but the wood used is on average at least 50 years old. The Indonesian-styled recycled

Recycling, HI-5.

> Producing aluminum by recycling takes 95% less energy than producing new aluminum form bauxite ore.
>
> —Courtesy of the EPA

teak products are strong and durable, and include day beds, dining sets, chairs, beds, and tables. The recyclable outdoor furniture has a more modern, angular style, but is a great choice for stylish outdoor furniture.

Another company pushing forward the element of recycling and sustainability in furniture and design is SoHa (www.sohaliving.com). They use reclaimed railroad ties to create unique bookcases, dining sets, and end tables. Often, the wood carries its own history that is incorporated into the "green" piece.

Whether the products are made from pre-consumer recycled products, material that is recycled before it is used by a consumer; postconsumer recycled product, a material or finished product that has served its intended use and then recovered before it is disposed of; or recovered materials, waste materials and byproducts recovered or diverted from solid waste landfills; there are plenty of choices for recycled products we can purchase to save energy required to make brand new products and conserve natural resources at the same time.

Here are a few products that are made with recycled products and are available to consumers: copier and printer paper, paper towels/toilet paper, envelopes, toner cartridges, scissors, corrugated containers, polystyrene peanuts, steel framing for construction, plastic lumber, roofing, wallboard, countertops, flooring, paint, clothes, re-refined motor oil, retread tires, used parts, mulch and compost, hoses, furniture, fencing. The list goes on and on, but this should pique your interest into the possibilities of what can be recycled and reused in another product.

Recycling reduces air and water pollution and emissions associated with landfilling and incineration, and conserves natural resources such as timber, water, and minerals by reducing the need for raw materials.

Cloth Diapers

Of course, the easy way to change a dirty diaper, wadding up the soiled plastic undergarment and tossing it in the trash, is a

---

www.eco-furniture.com is a great online resource for all types of eco-friendly furniture. Check out the eco-facts link for information on the company's ecological and environmental ethics, goals and causes.

Dolphin Diaper Service
P.O. Box 894538
Mililani, HI 96789
(808) 261-4775
dolphinds@hawaii.rr.com
www.dolphindiaperservice.com

commonplace, though environmentally harmful and wasteful practice. Over a billion trees are cut down every year to produce disposable diapers that take up to 500 years to biodegrade in a landfill.

A baby stays in diapers an average of 30 months; that's about 6,700 diapers per child taking up space in the landfills. The statistics—and the smell—are enough to make you cringe.

There is an alternative to disposable diapers that is gaining momentum among environmentally conscious families through out Hawai'i—cloth diapers. Cloth diapers can made from a handful of materials including 100% cotton, organic cotton, and hemp to name a few. They offer a sustainable solution to throwing away a handful of diapers everyday.

Cloth diapers are easy to clean: just rinse and throw them in the washer. When you wash your own cloth diapers, you ensure that no chemicals—like those sometimes found in disposable diapers—are used to treat the diapers. By using cloth diapers, you are saving the energy needed to produce all those disposables, and during that 30-month diaper stretch, you can save about $2,500. Line drying those tighty whiteys further increases your energy savings.

If washing dirty diapers doesn't sit well with your stomach, then let a diaper service do the dirty work for you. The Dolphin Diaper Service has been thinking green since 1990 and is a member of the National Association of Diaper Services, an organization that ensures quality and safety of members' products, processing, and service through careful monitoring of the laundering and sanitation practices. Dolphin Diaper Service picks up dirty diapers and delivers clean ones on a dependable schedule. You don't have to soak them or wash them first; just trade them out for clean diapers. They even offer new water-resistant diaper covers that replace pins and waterproof pants.

Eco-friendliness, not to mention the advantage of a diaper service, make cloth diapers a smart choice for a green home.

Moms Going Green is a grassroots organization dedicated to the health of families and the environment. Email Tara, co-founder of Moms Going Green, at tara@moms-goinggreen.com for more information about the group and "going green" workshops.

Locally owned Baby AWEARness sells different types of cloth diapers and other natural parenting products. They are located on O'ahu, but do online sales for the neighbor islands. www.babyawearness.com

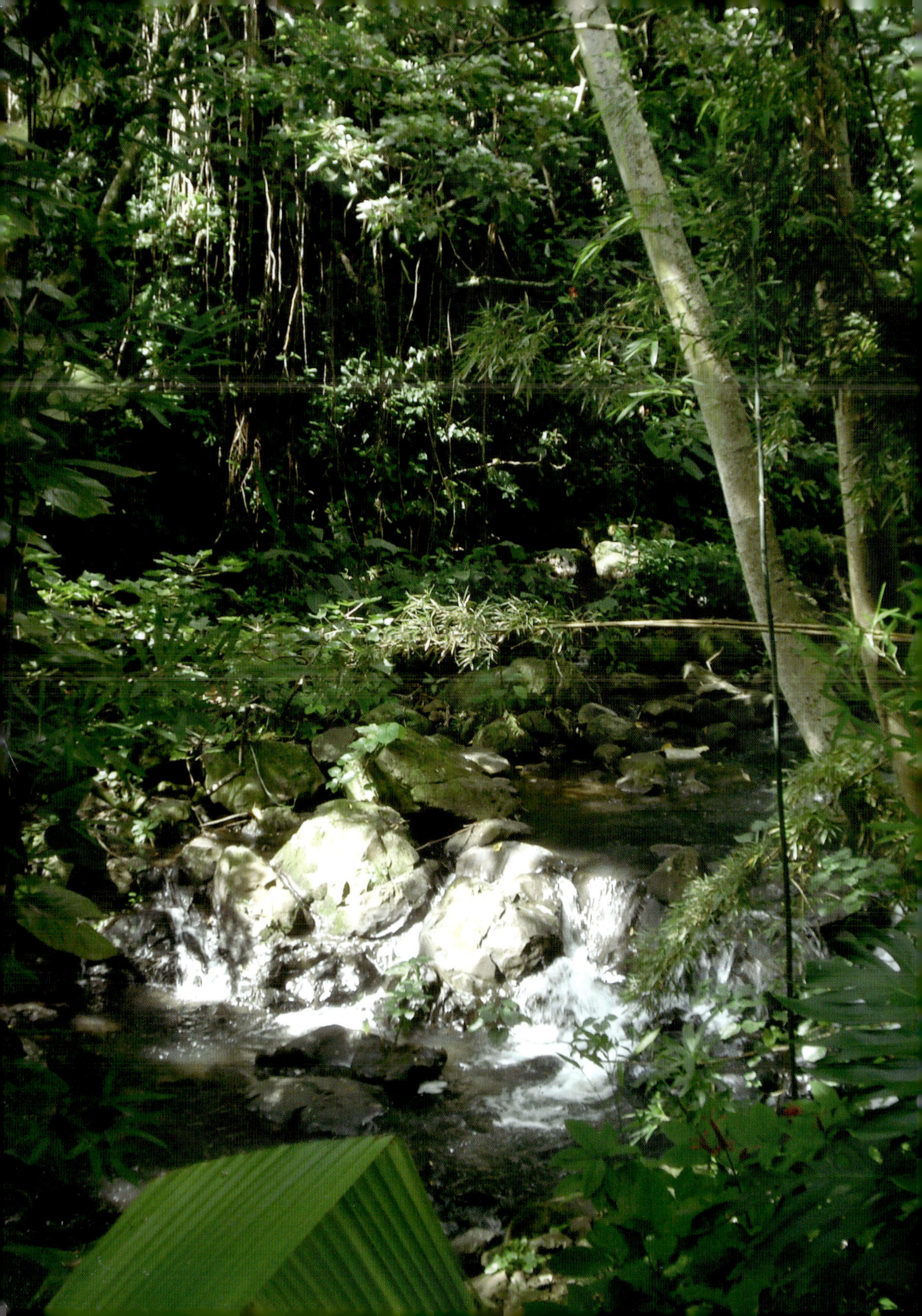

eight: volunteering

When you step out of home and into the community, you should have a sense of pride in your environment, your neighborhood, and your island—a true sense of place. And getting involved in your community through volunteering is one of the best ways to develop and share that pride and spread the ideology of your green home, perhaps laying the foundation for a greener community.

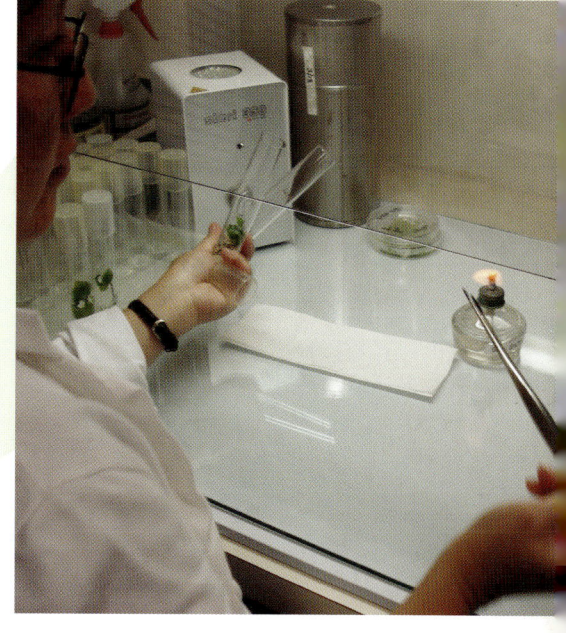

Lyon Arboretum micro-propagation lab volunteer.

Opportunities to volunteer for the betterment of your community abound; you just have to know where to start your search. The popular website Craigslist has a volunteer section for Hawai'i. Search for other opportunities online using terms like "volunteer" and the name of your island or town. Whether it's working with children, like in the Head Start program;

in the community

> Websites like volunteerhawaii.org are designed to help you search for volunteering opportunities.

working with animals, like volunteering at the local humane society or counting whales during their annual migration to the Hawaiian Islands; or even working with plants by volunteering at your local botanical garden and helping eradicate invasive species or working in a propagation lab, there are plenty of ways to get involved. These programs rely on people in the community to lend a hand.

Volunteer for an organization that shares the same ideals and morals as you, and then work together to achieve community goals. You'll do your best work when you're excited about what you are doing.

Maybe you can't find a group or an activity that seems to fit your interests. That's OK, too. It gives you the opportunity to start your own community program. It's always the right time for a beach clean-up or a weed-pulling makeover for an overgrown empty space—the keiki will definitely appreciate another place to play in the dirt.

Lyon Arboretum.

If you are ocean minded, the Surfrider Foundation combines a fun and relaxed atmosphere with a forum for the environmental watchdog. They're not afraid to tackle all manner of marine and coastal issues, from public beach access and beach clean-ups to fishing regulations and legal matters. And they still have time for a surf.

For environmental sorts, check out your local Sierra Club chapter.

If you have a green thumb, botanical gardens rely heavily on the help of volunteers for all sorts of tasks, from guiding tours to getting on your knees and getting your hands dirty.

Check out this online resource for nonprofit organizations: www.hawaii-county.com and search nonprofits for a link to the online resource guide.

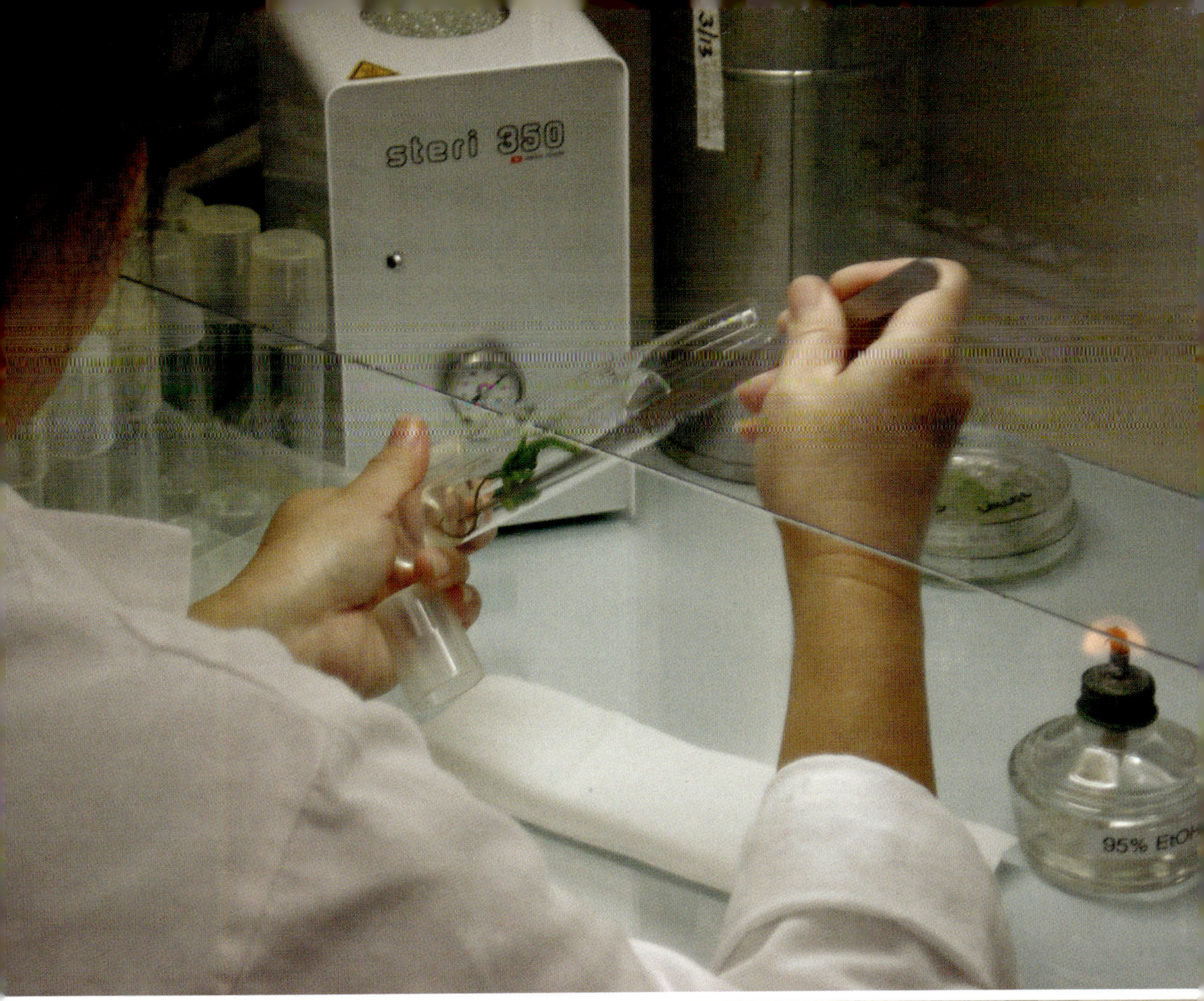
steri 350
95% EtOH

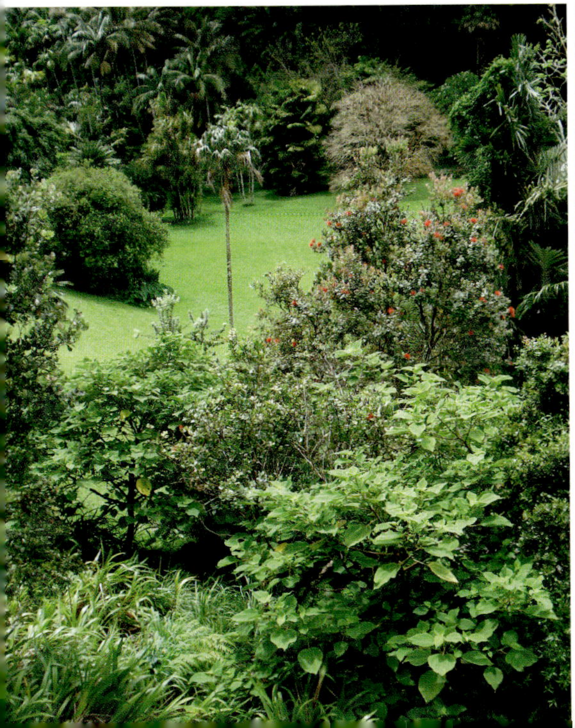

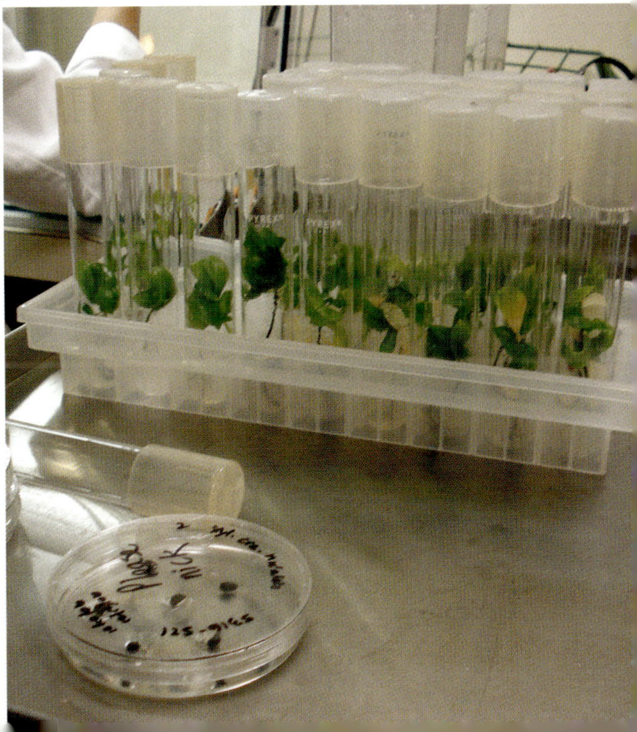

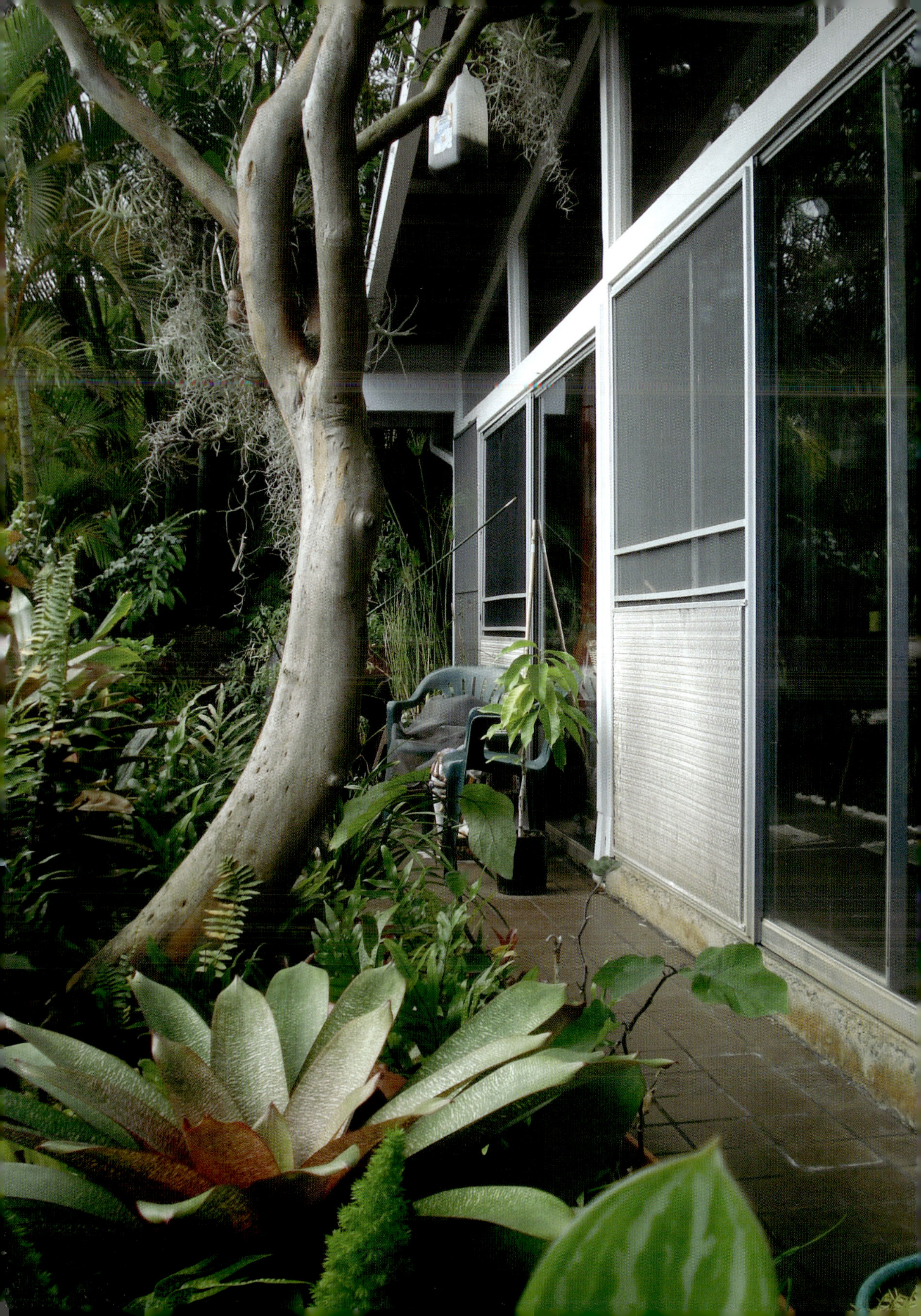

nine: comfortable living

During a three-month period in the last quarter of 2001, a monumental experiment was conducted on energy efficiency and comfortable living, "green" living, on the Hawaiian Homelands in the Waiʻanae area of Oʻahu. The American Institute of Architects Committee on the Environment, Honolulu Chapter, in partnership with the Department of Hawaiian Homelands, the Department of Energy, the State Energy Office, and the Utility and Building Industry Association undertook the project to quantify the effectiveness of passive-cooling design strategies and the resulting increase in comfort and reduction in energy demand on the Hawaiian home. They wanted to prove that it was possible to have a well-lit, comfortable, and cool home with maximum energy efficiency and monetary savings, all without air conditioning.

The "model" home.

in the community • 108

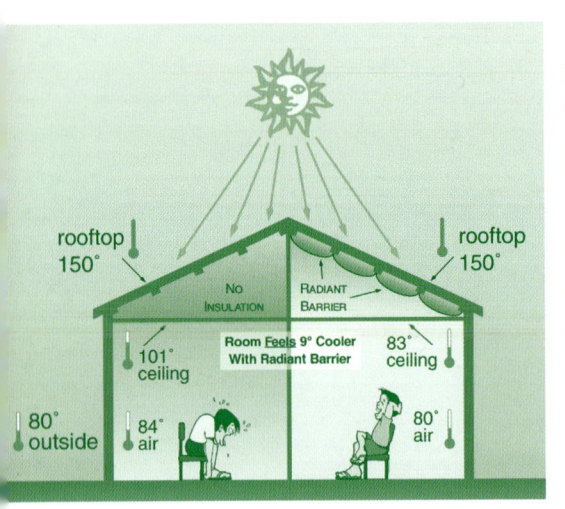

Radiant Barrier means no need for air conditioning; energy and money saved!

A "model" home was constructed just one block away from an identical existing home, known as the "control" house. The model home was built with energy-efficient and passive-cooling design strategies in mind, utilizing principles of natural ventilation. It contained a solar hot water heater; energy efficient appliances and lighting; radiant barriers in the roof and the south, east, and west walls; and a continuous ridge vent that ran the length of the roof. The exterior surfaces of the roof and walls were a lighter color to reduce heat absorption. The interior featured increased window openings, skylights, and ceiling fans. Neither house had insulation in the roof or walls.

Temperatures were measured externally at five interior locations in each home and data was taken on relative humidity, air movement, and illumination levels. As the roofs and walls of both houses heated up every day and logged the same temperatures, two very different scenarios were occurring inside the homes. While the temperatures in the attic of the control home during the hottest part of the day reached 26 degrees above the outside temperature, the model home's attic never raised 6 degrees above of the outside temperature. The radiant barrier effectively blocked the heat from entering the attic and the ridge vents allowed any heated air to escape through convective process at the highest part of the roof. This 20-degree difference in attic temperatures translated to a 9-degree difference in inside temperatures. That's right: the model home was as much as 9 degrees cooler than the control house with no radiant barrier.

Steven Meder, professor at the School of Architecture and Director for the Center of Smart Building and Communities, was an integral part of the experiment and a strong proponent for passive cooling design in Hawai'i. "In our location, the first passive design strategy is mitigating the heat that can get through the envelope. The section of the envelope that has the greatest impact is the roof because the sun is at a higher angle at our latitude. The second strategy is to evacuate heat by ventilation—any heat that has gotten in or been generated in the inside."

The second step according to Meder is to improve ventilation and airflow in the living space, a design utilized in the model house. "We had more cross ventilation that added to the comfort, although it didn't necessarily affect the temperature, but it evacuated some of the heat which makes people feel cooler because there is air moving across their body, which was not the case in the control house. Those passive design strategies got the building into the comfort range."

Not only was the model home comfortable and airy, but it was also designed to allow more natural light to filter through the interior of the house. The skylight was able to harness natural light and spread around indirect, ambient light, without raising the temperature of the living space. Sensors measured nearly five times more natural illumination in the model house.

The family living in the model home, utilizing their solar hot water, the energy efficient lighting and appliances, and passive cooling techniques, used about 40% less energy and saved over $650 that year in electricity bills, according to the final report released a year later. On a larger scale, families that decide to incorporate these principles into their home help Hawai'i reduce its dependence on imported oil and reduce emissions from power plants.

Passive design strategies are feasible and tangible, and make a lot of sense for out climate, our Islands, and our environment. They are simple strategies we can employ in our own homes, and this experiment shows the benefits across the board—environmental benefits, financial benefits, and quality-of-life benefits for you and your neighbors.

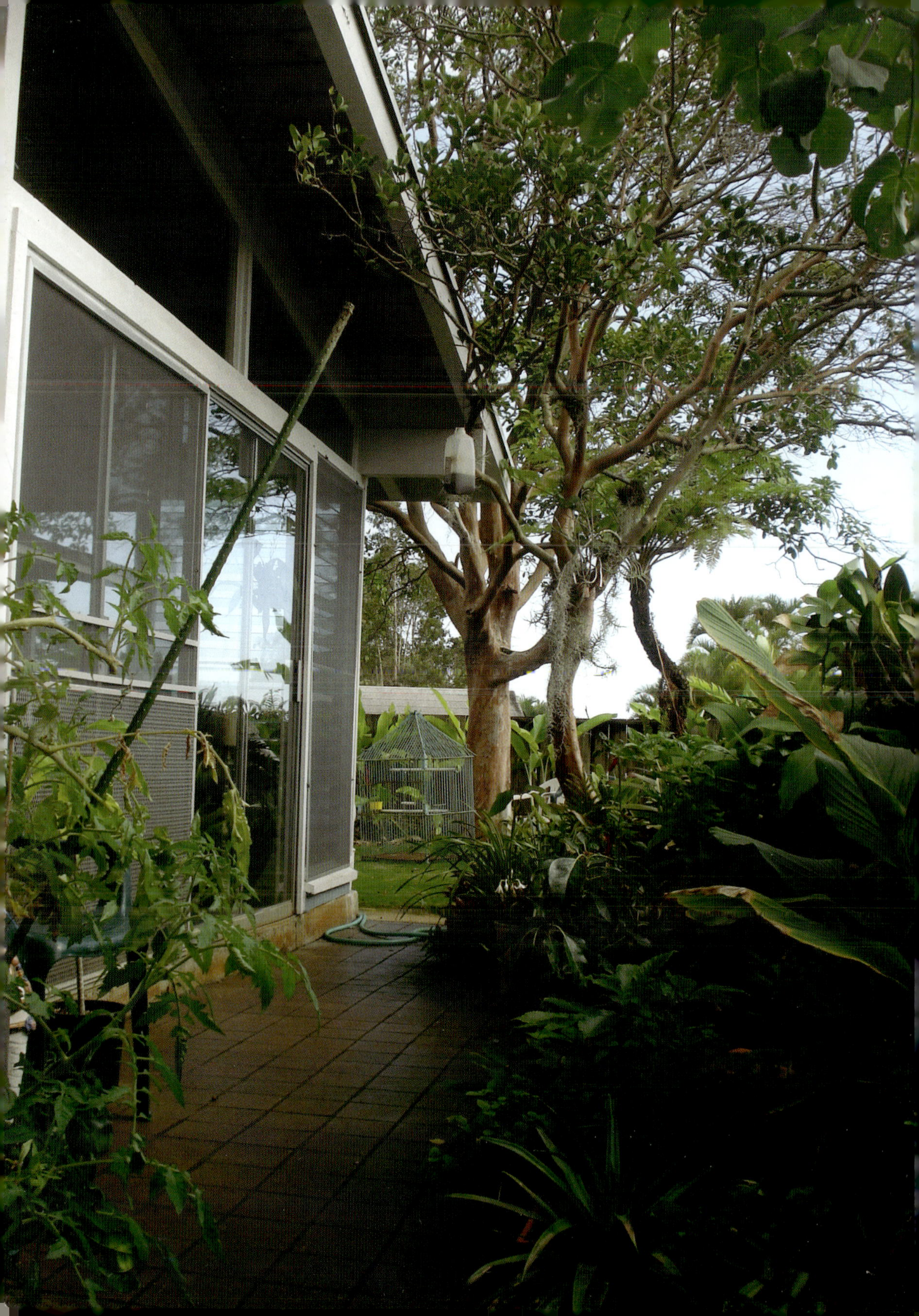

APPENDICES

References and Further Reading

Choi, Moom Yun. Bright Idea. *Honolulu Weekly*, November 1-7, 2006.

———. *Energy Tips & Choices: A Guide to an Energy-Efficient Home*. Honolulu: Hawaiian Electric Company, Inc., 2006

City and County of Honolulu Board of Water Supply. "Hawai'i Backyard Water Conservation: Ideas for every homeowner." www.boardofwatersupply.com/files/hawaiibackyardconservation.pdf.

Cuddigy, L. W., and C. P. Stone. *Alteration of Native Hawaiian Vegetation: Effects of humans, their activities and introductions* (pp. 103-107). Honolulu: The Pacific Cooperative Studies Unit and The Hawai'i-Pacific Island Cooperative Ecosystems Studies Unit, University of Hawai'i at Mānoa, 1990.

"Data Collection and Analysis of the Heat Mitigating, Passive Design Strategies at the Waianae Model Demonstration House." www.hawaii.gov/dbedt/info/energy/publications/modelhome2002.pdf.

Dinges, K., et al. *Composting Worms for Hawai'i*. Honolulu: College of Tropical Agriculture and Human Resources, University of Hawai'i at Mānoa, 2005.

———. *Small-Scale Vermicomposting*. Honolulu: College of Tropical Agriculture and Human Resources, University of Hawai'i at Mānoa.

Efficient Windows Collaborative. "Window Technologies: Low-E Coatings." www.efficientwindows.org/lowe.cfm.

Elevitch, Craig, and Wilkinson Kim. *Sheet Mulching: Greater Plant and Soil Health for Less Work*. Holualoa: Agroforestry Net, Inc., 1998

Grabowsky, Gail L. *50 Simple Things You Can Do To Save Hawai'i*. Honolulu: Bess Press, 2007.

Green Hawai'i. Published as a supplement to *Hawaii Home + Remodeling magazine*. Honolulu: Pacific Basin Communications, 2007.

Hill, Tiffany. "Gall Wasps Infect Native Plants." *Ka Leo*. October 24, 2006.

Kim, Karl. "Embracing Green Homes in Hawai'i." *Honolulu Advertiser*, Monday, December 10, 2007.

Macomber, Patricia S. H. *Guidelines on Rainwater Catchment Systems for Hawai'i*. Honolulu: College of Tropical Agriculture and Human Resources, University of Hawai'i at Mānoa, 2004.

"Milestones In Garbage." www.epa.gov/msw/timeline_alt.htm.

Pratt, Douglas H. *A Pocket Guide to Hawai'i's Trees and Shrubs*. Honolulu: Mutual Publishing, 1998.

Ritz, Mary Kaye. "Caring for Family and Mother Earth." *Honolulu Advertiser*, Sunday, January 6, 2008.

Schaefers, Allison. "Going Green." *Honolulu Star-Bulletin*, Sunday, May 27, 2007.

"Solar." www.hawaii.gov/dbedt/info/energy/renewable/solar/

Slayter, Mary Ellen. "What to Reuse and What to Buy New." *Honolulu Advertiser*, Sunday, January 6, 2008.

State of Hawai'i Department of Business, Economic Development and Tourism, Energy Resources and Technology Division. *Have Some Energy On The House ... Solar*. Honolulu: DBEDT, 2003.

———. *Hawai'i Homeowner's Guide to Energy, Comfort and Value*. http://www.hawaii.gov/dbedt/info/energy/publications/hhog.pdf.

———. *Hawai'i Recycling Industry Guide*. www.hawaii.gov/dbedt/info/energy/publications/recycling99.pdf.

U.S. Department of Energy. "A Consumer's Guide to Energy Efficiency and Renewable Energy." www.eere.energy.gov/consumer.

U.S. Department of Energy. "Radiant Barrier Attic Fact Sheet." www.ornl.gov/sci/roofs+walls/radiant/rb_01.html.

Surviving wiliwili trees: Koko Crater Botanical Garden.

About the Author

Kevin J. Whitton is the managing editor of *FreeSurf Magazine* on the north shore of Oʻahu and contributing writer to regional, national, and international publications. An avid naturalist and botanical hobbyist, he is also the author of *Rain, Beans and Rice: Memoirs of Life in a Costa Rican Rain Forest*, an adventure travel memoir based on his three-month experience as a trail guide in a private rainforest preserve. He has lived in California, Colorado, Mexico, and Australia, and now resides with his wife and two cats on Oʻahu.